THE SCIENTIFIC ENTERPRISE

BOSTON STUDIES IN THE PHILOSOPHY OF SCIENCE

VOLUME 146

THE SCIENTIFIC ENTERPRISE

The Bar-Hillel Colloquium: Studies in History, Philosophy, and Sociology of Science

Volume 4

Edited by

EDNA ULLMANN-MARGALIT

SPRINGER SCIENCE+BUSINESS MEDIA, B.V.

Library of Congress Cataloging-in-Publication Data

The Scientific enterprise / edited by Edna Ullmann-Margalit.
p. cm. -- (The Bar-Hillel colloquium ; v. 4) (Boston studies in the philosophy of science ; v- 146)
ISBN 978-0-7923-1992-4 ISBN 978-94-011-2688-5 (eBook)
DOI 10.1007/978-94-011-2688-5
1. Science--History--Congresses. 2. Science--Philosophy--Congresses. 3. Science--Social aspects--Congresses. I. Ullmann-Margalit, Edna. II. Series. III. Series: Boston studies in the philosophy of science ; v- 146.
Q124.6.S394 1992
509--dc20 92-31007

ISBN 978-0-7923-1992-4

Prepared in cooperation with Mrs. Esther Shashar, executive editor.
The Van Leer Jerusalem Institute.

THE BAR-HILLEL COLLOQUIUM: STUDIES IN HISTORY, PHILOSOPHY, AND SOCIOLOGY OF SCIENCE

The Israel Colloquium for the History, Philosophy, and Sociology of Science was established in the academic year 1981–82. It offers, annually, a series of public lectures, alternately in Jerusalem and Tel Aviv. It is sponsored and directed jointly by three bodies: The Center for the History and Philosophy of Science, Technology, and Medicine of the Hebrew University of Jerusalem; The Institute for the History and Philosophy of Science and Ideas of Tel Aviv University; and The Van Leer Jerusalem Institute. The advisory committee of the Israel Colloquium represents all the institutions of higher learning in Israel. In 1988 the Colloquium was renamed "The Bar-Hillel Colloquium."

Contents

Preface

The volume before us is the fourth in the series of proceedings of what used to be the Israel Colloquium for the History, Philosophy and Sociology of Science. This Colloquium has in the meantime been renamed. It now bears the name of Yehoshua Bar-Hillel (1915–1975). Bar-Hillel was an eminent philosopher of science, language, and cognition, as well as a fearless fighter for enlightenment and a passionate teacher who had a durable influence on Israeli philosophical life.

The essays collected in this volume have of course this much in common, that they are all in, of, and pertaining to science. They also share the property of having all been delivered before live, and often lively, audiences in Jerusalem and in Tel Aviv, in the years 1984–1986. As is customary in the volumes of this series, the essays and commentaries presented here are intended to strike a rather special balance between the disciplines to which the Colloquium is dedicated. The historical and sociological vantage point is addressed in Kramnick's and Mali's treatment of Priestley, in Vickers' and Feldhay's studies of the Renaissance occult, and in Warnke's and Barasch's work on the imagination. From a philosophical angle several concepts, all material to the methodology of science, are taken up: rule following, by Smart and Margalit; analysis, by Ackerman; explanation, by Taylor; and the role of mathematics in physics, by Lévy-Leblond and Pitowsky. In addition, the volume contains the proceedings of two symposia dedicated to two towering scientific figures: one celebrates Bohr's centennial, and the other examines "the other Newton."

I would like to note here, with sadness, the deaths of Shmuel Sambursky (1900–1990), and Shlomo Pines (1908–1990), who were both among the founding members of the Israel Colloquium. Both brought together in a splendid union the two worlds of science and humanities, Sambursky in his work on the science of Antiquity, and Pines in his contribution to Medieval scholarship as well as that of late Antiquity. With their deaths the world of learning lost two great scholars.

Edna Ullmann-Margalit

Eighteenth-Century Science and Radical Social Theory: The Case of Joseph Priestley's Scientific Liberalism

ISAAC KRAMNICK

In 1794 Joseph Priestley fled England and the Church-and-King sentiment that had set ablaze his Birmingham house and laboratory in 1791. His flight to America was noted by the United Irishmen with a public letter. This most radical group in the entire camp of English sympathizers with the French Revolution not only lamented English repression but offered a marvelous hymn to the tripartite linkage of America, useful science, and radical change.

> The emigration of Dr. Priestley will form a striking historical fact, by which alone future ages will learn to estimate truly the temper of the present times.... But be cheerful, dear Sir, you are going to a happier world, the world of Washington and Franklin. In idea we accompany you.... We also look to the new age when man shall become more precious than fine gold, and when his ambition shall be to subdue the elements, not to subjugate his fellow creatures, to make fire, water, earth, and air obey his bidding, but to leave the pure ethereal mind, as the sole thing in nature free and invincible.... The attention of a whole scientific people [here] is bent to multiplying the means and instruments of destruction ... but you are going to a country where science is turned to better use (Rutt 1831, 2:218).

The relationship between science and progressive politics was by no means one-way. Just as science would ameliorate the human condition, so a

E. Ullmann-Margalit (ed.) The Scientific Enterprise, 1–31.

progressive politics would encourage scientific advances. In the very last scientific paper he wrote before his self-imposed exile, Priestley made a digression into politics, noting the extent to which politics, ideology, and science were intertwined, and firmly identified himself and other radicals as the true virtuosi of science. In his *Experiments on the Generation of Air From Water* (1793) he notes

> Since ... the friends of philosophy in this century must separate on the ground of *religion* and *politics* ... may the separation have no farther consequences than that producing a generous emulation who shall most advance the cause of science; the friends of *Church* and *King*, as they affect exclusively to call themselves, or the friends of liberty, among whom, at all times, and in all circumstances, I shall be proud to rank myself. We are, it is true, but a small minority, but not deficient, I trust, in ability, activity, and energy, qualities which will always make men respected, though oppressed (quoted in Gibbs 1967, 243; emphases are Priestley's).

No surprise, then, that in his adopted America Priestley's champion would be the Founding Father who similarly linked progress to the fortunes of science. Jefferson wrote to Priestley on 21 March, 1801:

> Our countrymen have recovered from the alarm into which art and industry had thrown them; science and honesty are replaced on their high ground, and you my dear Sir, as their great apostle, are on its pinnacle. It is with heartfelt satisfaction that, in the first moments of my public action, I can hail you with welcome to our land, tender to you the homage of its respect and esteem, cover you under the protection of those laws which were made for the wise and good like you. Yours is one of the few lives precious to mankind for the continuance of which every thinking man is solicitous (quoted in Ford 1905, 9:216).[1]

Joseph Priestley was the preeminent scientist in the Anglo-American world in the era of the American and French revolutions. It is through his career and writings that the links in late eighteenth-century Britain between science, dissenting theology, and political radicalism can be investigated. This linkage will be described in both sociological and conceptual terms. The sociological dimension emphasizes Priestley's personal role as the central figure in political, educational, and religious radicalism in his native Britain. The conceptual dimension involves a discusssion of Priestley's Enlightenment vision of science as the redemptive agent of emancipation and amelioration, as well as his sense of the role of science in the demystification of the world and the polity. Also noted will be Priestley's use of the scientific method as a metaphorical model of unfolding truth and perfection, with the American and French revolutions as decisive real-world "experiments." Finally, this paper will investigate Priestley's psychology, with its mechanistic image of the mind, which, in turn, is central to his scientific image of society and its potential for being easily and dramatically reformed. Before turning to this examination of Priestley's scientific liberalism, however, his remarkable career deserves some notice.

I. "One of the Few Lives Precious to Mankind"

Known to later ages as the great scientist who discovered oxygen, or perhaps as the theologian who founded modern unitarianism, to his contemporaries Priestley was known primarily as "Gunpowder Joe," the leading radical intellectual of his era. In the 1770s Priestley was, along with Richard Price, one of the most ardent supporters of the American cause. This outspoken support for the Americans was, in fact, soundly rebuked by Samuel Johnson in the latter's famous attack on the colonists in 1775. Priestley's marriage of science and radical politics was not to Johnson's taste. "Ah, Priestley," Johnson is reported to have uttered, "an evil man, Sir. His work unsettles everything" (quoted in Taylor 1954, 11; see also Johnson 1775). More significant for Priestley's "evil" reputation was his emergence in the 1780s as the leader of the English reformers, who called for a reform of Parliament, the extension of the suffrage, and the repeal of the onerous Test and Corporation Acts, which required students at Oxford and Cambridge, as well as all holders of governmental and municipal positions, to be communicants of the Church of England. To all of this Edmund Burke replied, "If I must, my choice is made. I will have rather George III or IV than Dr. Priestley" (Burke 1791–92, 1389).

Priestley's pamphlets called not only for an end to the legal disabilities faced by dissenters like himself and his fellow unitarians, but for disestablishment of the Anglican church and total separation of church and state. To no avail, however, for in 1787, 1789, and 1790, Parliament refused to repeal the acts. In the last of these efforts, Burke led the forces for the status quo. To repeal these fundamental religious laws, Burke warned, would "proceed step by step" to a leveling of the "foundations" of first the church and then the state (Burke 1791–92, 1387).

That Burke could argue so reflected the pervasive popular linkage of the religious rebel Priestley with the cause of political radicalism. The assault on the Test and Corporation Acts did, indeed, come from the very same dissenting intellectuals and industrialists who were demanding a reform of the archaic Parliament, which saw newly populated Midlands cities like Birmingham, Manchester, and Leeds unrepresented while rural constituencies of fields and cows continued to send members to Parliament. What frightened Burke and others, of course, was that the assault on the Test and Corporation Acts and the demands for political reform were intensified following the news of the revolution in France; this provided the background for the mob's burning of Priestley's house in 1791.

Priestley was lyrical in his praise of the French Revolution, as were virtually all the dissenters. For Priestley, the revolution was "unparalleled in all history," an event ushering in "a most wonderful and important era in the history of mankind." It would transform human existence "from superstition to sound knowledge," "from a most debasing servitude to a state of the most exalted freedom." The French and American revolutions were described by Priestley in a language that sounds strikingly like the later Marx. They represented

> a liberating of all the powers of man from that variety of fetters by which they have hitherto been held. So that, in comparison with what had been, now only can we expect to see what men really are, and what they can do (Priestley 1791, Letter XIV, 143–44).

Burke's *Reflections on the Revolution in France*, published in November of 1790, singled out Priestley and Price as the leaders of a dissenting plot to subvert the English social order as the French had subverted theirs. There was no mistaking Burke's placing Priestley at the head of this pack of seditious malcontents when, describing their conspiracy against the established order, he noted that "the wild gas, the fixed air is plainly broke loose" (Burke 1968, 90). Priestley immediately replied in a pamphlet attacking Burke's *Reflections*. There was little restraint in his rebuttal.

> Your whole book, Sir, is little else than a vehicle for the same poison, inculcating, but inconsistently enough, a respect for princes, independent of their being originally the choice of the people, as if they had some natural and indefeasible right to reign over us, they being born to command, and we to obey (Priestley 1791, Letter III, 30).

On 14 July 1791, many dissenters in Birmingham gathered at the Temple Row Hotel to commemorate the French Revolution. The call to the meeting urged that "surely no free-born Englishman can refrain from exulting in this addition to the general mass of human happiness. It is the cause of Humanity!

It is the cause of the People!" (Gibbs 1967, 199). The people of Birmingham, alas, were not moved. Inflamed by Anglican preachers, as well as by free liquor, the mob that evening destroyed the dissenter chapel and headed on to Priestley's house and laboratory. Priestley was warned and fled; but he had no time to take anything with him. His house was burned to the ground, and with it his manuscripts, library, and scientific equipment. For days the mob destroyed houses of other known dissenters and radicals. Finally, Dundas, the home secretary, sent royal troops to restore order. Priestley had traveled through the night to London. George III took note of him in authorizing his secretary of state to send troops to Birmingham. The king wrote, "I cannot but feel better pleased that Priestley is the sufferer for the doctrines he and his party have instilled, and that the people see them in their true light" (Dobree 1935, 212).

In 1794 Priestley sailed for America. Samuel Taylor Coleridge, still in his radical phase, wrote of this shame (in his *Religious Musings* [December 1794], 371, cited in Willey 1961, 195):

> Him from his native land,
> Statesman blood-stained and priests idolatrous,
> By dark lies maddening the blind multitude,
> Drove with vain hate; calm, pitying, he retired,
> And mused expectant on these promised years.

In America Priestley would settle in rural Pennsylvania, at the fork of the Susquehanna River, near Northumberland, where Coleridge and the poet Southey had once dreamed of creating the utopian community they called Pantisocracy. Here Priestley would live until his death in 1804 and here the American Chemical Society would be founded in 1874, on the centennial of Priestley's discovery of oxygen. Nothing pleased Priestley more during his last ten years in America than Jefferson's election to the presidency in 1800. It marked a personal milestone for Priestley. He wrote, "I now for the first time in my life (and I shall soon enter my 70th year), find myself in any degree of favour with the government of the country in which I have lived, and I hope I shall die in the same pleasing situation" (quoted in Holt 1931, 205).

II. Science and the Liberal State

Priestley's life, then, was itself a testimony to radical politics in the late eighteenth-century Anglo-American world. His commitment to social change is best seen, however, in his writings, where he grounds that reformist impulse in his scientific world view. At the base of his social theory is scientific conviction.

In Priestley and his circle, the English had their eighteenth-century *philosophes* and their Enlightenment. They produced no *Encyclopaedia*, but their impact and influence were every bit as great as those of their French counterparts. As Priestley sought to free religion from superstition, so that it could be accepted by "philosophical and thinking persons," so his writings, politics, and science were permeated with an optimistic and progressive outlook. The advancement of science in the late eighteenth century was based on the nexus in England of rising industry, political reform, and liberal theology. At the center of this nexus stood Priestley (Rutt 1831, 75; see also Bernal 1965, 481).

The quest for happiness on this earth preoccupied Priestley the *philosophe*. It informed his religion and it structured his science. In his *Catechism for Children and Young Persons* (pp. x, 27), the question is posed "What did God make you and all your mankind for?" The answer is given: "He made us to be good and happy." Even more significant is the reply to the question, "Will not an application to worldly business interfere with the duties of religion?" "No," Priestley replies, "we please God the most, by doing that which makes ourselves and others the most happy." In his *Sermons* (p. 190) the same theme is struck. Our lives and work have one objective: "the glorious, animating prospect of the happy state of mankind."

Priestley was convinced that the religious life was a life of temporal happiness, one most easily attainable by the hard-working bourgeoisie to whom he preached and for whom he wrote. When he looked back on his years near the great aristocrats in England, like Lord Shelburne, for whom he had once worked, he used this theme of happiness to symbolize his discontent with the great.

> I was not at all fascinated with that mode of life.... The people are generally unhappy from the want of necessary employment on which accounts chiefly there appears to be much more happiness in the middle class of life, who are above the fear of want, and yet have a sufficient motive for constant exertion of their faculties, and who have always some other object besides amusement (*Memoirs*, 82).

The great in the aristocratic houses, however, were not sufficiently exerting their faculties on "carrying on improvements," which was, according to Priestley, along with the "remedy of abuses," the mission God had given humanity through his injunction that it be useful through "industry and zeal" (*Proper Objects of Education*, 2, 6; *Sermons*, 21). Science was, of course, the "necessary employment" most central to increasing happiness, remedying abuses, and furthering progress. In the preface to *The History and Present*

State of Electricity (1769), Priestley invokes Bacon. Knowledge becomes power as the scientist "is master of the powers of nature." This power is applied "to all the useful purposes of life." The scientist contributes to the "security and happiness of mankind." Less interested in unlocking secrets of the universe, or in knowing God's ways, the scientist has the practical goal of making "human life ... in its present state ... more comfortable and happy ... able to subsist with more ease." This is, indeed, a civic responsibility, for by his work the scientist serves the community. He is "a good citizen and a useful member of society" (*History of Electricity,* pp. iv, 442).

The practical and political aims of Priestley's science were most persuasively articulated by his most important disciple, Thomas Cooper. A fascinating man in his own right, Cooper was a Manchester manufacturer and chemist, and a political radical who became Priestley's closest philosophical and religious follower. He wrote numerous essays on materialism and unitarianism both in England and in America, where he also fled in the 1790s. He would also edit much of Priestley's work, especially the latter's *Memoirs*, which he in fact completed. Cooper would ultimately become professor of chemistry at Columbia, South Carolina, and president of the University (for details on Cooper's career, see Malone 1961). Cooper developed Priestley's philosophy of science even more self-consciously. Bacon was indeed the inspiration for the English as well as the French *philosophes*. Bacon's genius was, according to Cooper, in turning scientific man away from the games of the learned, "playing tricks with syllogisms, and the legerdemain of words." He turned man to experiments and facts, to practical, ameliorative concerns. Chemistry was central in this practical turn. "Chemistry is of more immediate and useful application to the everyday concerns of life and it operates more upon our hourly comforts than any other branch of knowledge whatever." Alongside chemistry, the crucial science, according to Cooper, was mechanics, the study of the laws of inorganic bodies, their size, shape, weight, forces, and properties. As cultivated in England, mechanics had contributed to the riches of the nation and the comfort of all its inhabitants. "It is to this science in conjunction with chemistry that the great wealth and power of that nation is principally indebted" (Cooper 1812, 7).

Science was, according to Cooper, the most praiseworthy pursuit of any would-be cultured man. Science was in fact the crowning glory of an advanced civilization, replacing literature, which was always suspect for the practical and ascetic dissenter, since it distracted attention from more productive things. The praise of science is also linked to middle-class disdain for the past, for history, and for custom. For the middle class, the mystery and awe of ancient institutions and ancient ideas were part of the barbaric feudal past, to be replaced by usefulness and efficiency. All that Burke found so comforting and so necessary for civilized existence, Cooper swept away in his description of science's contribution to the progress of man.

> What should make the knowledge of those laws on which all useful manufactures depend less interesting than a knowledge of the licentious poets and dramatists? It is in the infancy of society chiefly that works of mere fiction receive an inordinate share of public attention. It is in our boyish years that the poets and novelists, the writers who are disgracefully employed in furnishing stimulus to appetites that require to be bridled, engage our attention. When experience has taught us wisdom, we begin to estimate utility as the criterion for desert, and look back with some regret at the time misemployed in mere amusement. It is so with the progress of civilized society; in the infancy and the ignorance of all communities, the great objects of intellect in peace are poetry and oratory; as nations advance in knowledge, science gains a rightful ascendancy (Cooper 1812, 13–14).

Priestley's and Cooper's science is a science of material and physical domination; it is "a knowledge that multiplies a thousandfold the physical force of a human being." It is a utilitarian and practical science relatively unconcerned with ultimate theories and speculative systems, but concerned with "rendering every hour of existence more desirable." Few have better stated the Enlightenment's conception of science than Priestley's disciple Cooper, as he captures all the progressive and ameliorative overtones of the eighteenth-century worship of science.

> Science compels every object around us to contribute, in some way or other, to our pleasure, to our profit, to our comfort, or to our convenience ... which multiplies not only human enjoyment, and alleviates human suffering, but multiplies also the human species; by providing more extensively the means of constant employment, and comfortable subsistence (ibid., 99).

All the while, of course, Cooper has the more immediate, materialist tribute to pay to science. "It is to science, chemical and mechanical, that England is indebted for having made her island the storehouse of the world, for having compelled the nations of the earth to pour into her lap their superfluous wealth, for having acquired the undisputed command of the sea.... Her merchants are as princes" (ibid., 96). Through science man compels objects, and nations compel other nations. Bacon has come a long way. In Priestley's circle, science is indeed power — the power of profits to transform merchants into would-be princes.

The model scientist for Priestley's circle was Benjamin Franklin. Here, in fact, one found the scientist as statesman-prince. And here was also the preacher of pleasure, profit, comfort, and convenience. Franklin and Priestley were, of course, close friends, who corresponded often about politics and electricity, and Franklin, we know, was a religious freethinker as well. A veritable cult of Franklin emerged in Priestley's circle. Many a hero in the early children's books produced by writers in Priestley's circle were named Benjamin. Indeed, Anna Barbauld, the radical disciple of Priestley who wrote poetry singing his praises and tracts defending his calls for the repeal of the Test and Corporation Acts, offered Franklin as the model of "true heroism" in the new age. In one of her stories for children she wrote:

> Few wiser men have ever existed than the late Dr. Franklin. His favorite purpose was to turn everything to use, to extract some potential advantage from his speculations. He understood common life and all that conduces into its comfort. He left treasures of domestic wisdom that were superior to any of the boasted maxims of antiquity (Barbauld 1792–96, VI:250; see also Kramnick 1980).

Like Franklin's, Priestley's science was a part of the newly organized and interlocking world of politics and economics at whose center he stood. Priestley got to the center of that world partly by good fortune. He had married one Mary Wilkinson, the daughter of Isaac Wilkinson, ironmaster. While her father was poor, Priestley writes, "I had a little fortune with her. I unexpectedly found a great resource in her two brothers, who had become wealthy, especially the elder of them." This brother-in-law, John Wilkinson, the founder of the Wilkinson steel fortune, introduced Priestley to his own wider circle of dissenter industrialists in the Midlands. As a result, Priestley became lifelong friends with Bentley and Josiah Wedgwood. These men and Wilkinson would be financial patrons of Priestley's scientific work throughout their lives. In turn, he would concern himself with the practical problems of industrial physics and chemistry. Also, they and their political, religious, and social concerns often turned Priestley's pen to the world of affairs.

The proliferating learned societies in provincial Britain, like the philosophical societies in Manchester, Derby, and Sheffield, and especially the Lunar Society in Birmingham, saw industrialists like Boulton, Wedgwood, and Watt and scientists like Priestley, Galton, and Darwin exchanging papers on the latest chemical breakthroughs, on electricity, or on theories of free trade and republican government. These societies bound science ever closer to the new world of profit and power, the world that Burke would characterize as that of sophistry, economy and calculation (for the

Lunar Society of Birmingham see Schofield 1963; see also Robinson 1953; 1956; Taylor 1943; Marshall 1946; Chaloner 1958).

This science was closely linked to radical politics. Its practitioners were themselves for the most part men of talent who, in a closed aristocratic society of privilege and rank, sought radical social changes, not the least of which was greater social and political rewards and power for themselves and their industrial friends. But science was also a powerful tool in the effort to demystify and simplify the universe. Here, too, its impact was radical. The ancien régime and the aristocratic political world were defended by Burke and others because of their very mysterious and superstitious essences — the dark, shadowy eminences of kings, queens, and lords with their cloaks of mysterious authority, crowns, scepters, and thrones. Government, it was held, was also a mysterious, complicated, and arcane realm. Only those born to it could understand and manipulate it. Science, which had expelled superstition from the heavens, could expel the mysteries that lay heavy on aristocratic society. Its new and corrosive ideals were truth, efficiency, and utility.

The practitioners of this science saw it as the handmaiden of a radical politics. In his *Experiments and Observations on Different Kinds of Air* (1774), Priestley was so convinced that science undermined all "undue and usurped authority" that he envisioned the English hierarchy with "reason to tremble even at an air pump or an electrical machine" (ibid., p. xiv). It was a natural reflex in Priestley's circle to link scientific advances with political reform. So it was when Wedgwood replied to a letter from his partner describing some recent work of Priestley's:

> I am much pleased with your disquisition upon the capabilities of electricity, and should be glad to contribute in any way you can point out to me towards rendering Doctor Priestley's very ingenious experiments more extensively useful.... But what daring mortals you are to rob the Thunder of his bolts — and for what? — no doubt to blast the oppressors of the poor and needy, or to execute some public piece of justice in the most tremendous and conspicuous manner, that shall make the great ones of the earth tremble! (Wedgwood to T. Bentley, 9 October 1776, in Wedgwood 1774, 44).

Priestley, in turn, used similar apocalyptic scientific imagery to describe the impending and irresistible reform of the antiquated religious-political order in Britain. It would be, in fact, this passage that would earn him his nickname "Gunpowder Joe." "We are," he wrote,

> as it were, laying gunpowder, grain by grain, under the old building of error and superstition, which a single spark may hereafter inflame, so as to produce an instantaneous explosion; in consequence of which that edifice, the erection of which has been the work of ages, may be overturned in a moment and so effectually as that same foundation can never be built again (*Reflections*, 101).

According to Priestley, science would "be the means, under God, of extirpating all error and prejudice, and of putting an end to all undue and usurped authority...." Science also inspired a millennial fervor in these dissenting circles. It was the basis for their unbounded faith in progress, their belief in perfectibility and the imminent elimination of pain and suffering. Priestley assumed that science would create a people "more easy and comfortable," who would "grow daily more happy." "Whatever was the beginning of this world," he wrote, "the end will be glorious and paradisiacal, beyond what our imaginations can now conceive" (cited in Gibbs 1967, 62; see also *Writings,* 198). Science might even eliminate mortality. So Franklin wrote to Priestley:

> The rapid progress the sciences now make, occasions my regrets sometimes that I was born so soon. It is impossible to imagine the heights to which may be carried in a thousand years the power of man over matter.... All diseases may by sure means be prevented or cured, not excepting even that of old age, and our lives lengthened at pleasure, even beyond the antediluvian standard (Rutt 1831, 1:329).

The institution critical for science, for the increasing amelioration of human life and for ushering in the millennium, was the dissenting academy, certainly not Oxford and Cambridge. Priestley was "sorry to see the ample revenues of the two universities ... applied in a manner so little favorable to the real interests of science." The source of progress was, on the contrary, the dissenting academy, where Priestley put "modern scientific subjects" at the center of the curriculum. He wrote to Pitt in 1787, "While your universities resemble pools of stagnant water secured by dams and mounds, and offensive to the neighborhood, ours are like rivers, which while taking their natural course, fertilize a whole country" (*Letter to Pitt*, 20). The dissenting academies served and by no means offended their different neighbors. Located primarily in the newer industrial communities and funded by dissenting industrialists, the academies taught subjects fit, as Priestley put it, "for civil and active life." The dissenting academy, he wrote, would break down the traditional barrier between "the learned world and the common world." Men trained for an active life would no longer receive a classical education. Science and mathematics would replace ancient languages or classical thought and criticism. "Those times of revived antiquity," Priestley wrote, "have had their use and are no more.... Their maxims of life will not suit the world as it is at present." What he did teach his students was accounting, algebra, geometry, modern history, commerce, theories of manufacture, and politics. Even those studying to be ministers, he insisted, should study mathematics and science. The purpose of education was, after all, "to find expedients to remedy abuses and carry on improvements." The clergy were being trained for that same crusade (*Writings*, 296; *Proper Objects of Education*, 6; for background on dissenting academies, see Parker 1914; McLachlan 1931, 1943).

The graduates of these academies industriously working in "civil and active life" were modernizing saints ushering in the social and political millennium. They would join the pantheon of Priestley's heroes, would become "a Bacon or Newton, a Locke or Hartley, a Hampden or Sydney, a William Penn, or Franklin, a Washington" (*Proper Objects of Education*, 5). These graduates brought closer a wondrous and dazzling secular age with governments confining "their attention to the civil concerns . . . and consulting their welfare in the present state only." As a result all people would soon be "flourishing and happy." All animosity between nations would come to an end and there would be universal peace and goodwill among nations. Priestley suggests that "when the affairs of the various societies shall be conducted by those who shall truly represent them," colonial empires would be no more; there would no longer be one people held in subjection by another. "The idle pageantry of a Court" would be replaced by canals, bridges, roads, libraries, and laboratories. This "happy state of things," when wars would end and men would beat their swords into plowshares, was, Priestley noted, repeatedly foretold in the prophecies of 2,000 years earlier. But there was more than Providence behind it, for it is "a state of things which good sense, and the prevailing spirit of commerce" worked to bring about. In "this new condition of the world," there would be no men singled out by titles, "a profanation of epithets." The magistrates would be "appointed and paid for the conservation of order." Government would be simple in scope and "unspeakably less expensive than it is at present." The taxes now paid, "a most unreasonable sum ... perhaps the amount of one half of our property," would "become superfluous" with the elimination of the state church and the aristocratic establishment (*Letters to Burke*, Letter XIV, 143–55; for Priestley's millennial thought, see Fruchtman 1983).

Millennial fervor was endemic among English intellectuals in the last decades of the eighteenth century. Wordsworth's "dawn when it was such bliss to be alive" was but the metaphorical symbol of a widespread belief in progress and perfectibility. Priestley, too, saw "the morning opening upon us, and we cannot doubt but that, the light will increase, and extend itself more and more unto the perfect day." Science was, of course, crucial. "As all things (and particularly whatever depends upon science) have of late years been in a quicker progress towards perfection than ever; we may safely conclude the same with respect to any political state now in being" (*History of Corruptions*, 4).

Nearly two centuries later we remember the confident optimism of the young Wordsworth and of those in Priestley's circle as they saw the progressive and ameliorative impact of science reflected in the revolutionary ideals of the American and French revolutions. It stays with us all the more because of the pathetic disillusion that would soon set in on such men as Wordsworth and Coleridge, though, it should be noted, never on Priestley. What we seldom remember, however, is the towering intellectual influence that informed and shaped the optimism of Priestley and his circle, that of David Hartley. Physician and Christian philosopher, Hartley, in that characteristic eighteenth-century English way, combined scientific zeal with religious certainty. He was the author in 1749 of *Observations on Man, His Frame, His Duty, and His Expectations*, the holy book for millenarian perfectionism. Praise was tendered Hartley in different ways. Coleridge named his first son David Hartley, students in the 1790s at the dissenting academy of Hackney, such as Hazlitt, read "Hartley for four and a half hours a week," and Joseph Priestley reprinted Hartley's book, noting, "I think myself more indebted to this one treatise, than to all the books I ever read beside, the Scriptures excepted" (quoted in Ashley Smith 1954, 177; *Examination*, p. xix).

Hartley's psychological theory of associationism provided the basis for his tremendous influence. "Dr. Hartley," Priestley wrote, "has thrown more useful light upon the theory of the mind than Newton did upon the theory of the natural world" (ibid., p. xv). Building upon Hobbesian and Lockean sensationalism and its insistence that the source of knowledge is found in the sensations caused by external objects acting on the senses, Hartley argued that ideas originating in this way and developing through physiological processes in the brain become "associated" together in a certain necessary order in the mind. The mind was itself a natural mechanism, governed by scientific laws, and these laws were in turn discernible through observation and provided a materialist base for a theory of human learning. Its relevance for educational doctrine accounts for its impact on Priestley and explains the linkage of materialism with optimism. For Priestley and others, like Jefferson and Rush, "the most important application of Dr. Hartley's doctrine of the association of ideas is to the *conduct* of human life, and especially the business of education" (ibid. p. xiii; see also D'Elia 1970). Men's minds were formed by their circumstances. The actions of the environment and of teachers had a necessary effect on children that could not be otherwise, argued necessitarian deniers of free will like Priestley. It was through education, then, that the millennium would be reached. John Stuart Mill understood this well as he reflected on his father's schooling in Edinburgh, a center of Hartleyan medical/psychological materialism in the 1790s. Hartley's influence convinced his father and others, he wrote, of "the formation of all human character by circumstances, through the universal principle of Association, and the consequent unlimited possibility of improving the moral and intellectual condition of mankind by education" (Mill 1924, 91).

Hartley, without the evidence of the American or French revolution, read the history of mankind as one of inevitable progress. From its infancy, mankind, with its "mind as a blank, void of ideas, as children now are born," had grown through the process of the association of ideas to an adulthood where its knowledge and happiness would inevitably be paradisiacal. Presiding over the workings of the association principle was God and his providentially designed "system of benevolence," which would lead individuals in society "to promote the welfare of others." This moral sense was not innate, as the Scottish philosophers argued, but "generated necessarily and mechanically" (Hartley 1749, 2:139; ibid., 1:473, 368, 510; for a discussion of Hartley, see Willey 1961).

Inherent in Harley's "optimistic materialism" and in that of his disciple Priestley is, of course, a paradox. As necessitarians, they claimed mankind to be the passive product of circumstances; as reformers, they preached active intervention in controlling and changing circumstances, in educating, in order to produce progress, perfection, and paradise. It is this latter perspective which most characterizes Priestley and his fellow Hartleyan reformers. Other close friends of Priestley who were also taken with Hartley's associationist theories were busy reforming all of England's institutions. Like Priestley, they were Protestant dissenters fueled by a zeal to improve and reform. John Howard was developing the principles of the modern penitentiary (see Ignatief 1978), Thomas Percival of Manchester was developing the first modern hospital (see Percival 1803), and Josiah Wedgwood was developing in his factories the first modern principles of scientific management (see McKendrick 1961). Priestley approached the polity, John Howard prisons, Thomas Percival hospitals, and Wedgwood factories, as problems of mechanics, where active intervention through scientific manipulation of circumstances — in other words, education — could produce "improvements," by necessarily altering the causal chain of associations in the machine which was man's mind. Warrington, Hackney, and other dissenting academies were thus the great agencies of redemption. Their commitment to science and to the psychology of Hartley laid the basis for what they thought would be the perfecting of England through education.

How appropriate, then, that Priestley's millennial vision of an ideal, comfortable, and happy world produced by science and political reform is best expressed in a lecture to the parents and supporters of the Academy at Hackney in 1791. It is a priceless monument to the progressive, optimistic, and utopian moment which Priestley represents in the history of science and liberalism.

> Train our youth to the new light which is now almost everywhere bursting out in favor of the civil rights of men, and the great objects and uses of civil government. While so favorable a wind is abroad, let every young mind expand itself, catch the rising gale, and partake of the glorious enthusiasm, the great objects of which are the flourishing state of science, arts, manufactures, and commerce, the extinction of wars, with the calamities incident to mankind for them, the abolishing of all useless distinctions, which were an off-spring of a barbarous age (producing an absurd haughtiness in some, and a base servility in others) and a general release from all such taxes and burdens of every kind, as the public good does not require. In short, to make government as beneficial and as little expensive and burdensome as possible.... (*Proper Objects of Education*, 22, 39).

III. Liberalism and the Scientific State

Priestley's scientific world view is at the core of his theory of the liberal state as well. To be sure, he begins with conventional contractual convictions; he was thus a Lockean in more than merely his materialist psychology. He was, as his enemies never tired of noting, "the most eminent disciple" of the Lockean contractual natural rights school (Kramnick 1982). Priestley's *Essay on the First Principles of Government* (1791) is a virtual gloss on Locke's *Second Treatise*. He begins by describing a state of nature filled with people "living independent and unconnected," who "voluntarily resign some part of their natural liberty" to magistrates who are then the people's servants. These magistrates are accountable to the people, and if they abuse the trust given them, they can be deposed. Punishing their governors is a natural right of the people. Priestley dismissed out of hand those who cited historical rights, specifically lost Saxon rights. Saxon England was a time of idleness, treachery, cruelty, and insecurity of property, Priestley argued. It was no golden age or moment of pure principle that should be returned to (*Essay on Government*, 2–7; *Lectures on History and General Policy*, 349).

In 1769, Priestley described government as set up to secure the rights of "my life, my liberty, my property, and my ease" (*View*, 31). In 1789, writing of Anglicans, dissenters, Catholics, and Negroes, he insisted that there were "common rights of humanity." All the variety of Englishmen were "equally men ... and therefore equally entitled to all the natural and just rights of man." To protect these rights, civil government was established. But, in good Lockean terms, Priestley emphasized the contractual nature of civil power. Governors were mere servants and "if such magistrates abuse their trust, in the people therefore lies the right of deposing and consequently of punishing them" (*Conduct*, 14; *Writings*, 206).

Priestley's fundamental maxim of politics was the need to limit state interference in individual liberty. There was, he wrote, a spirit of the age which "encourages us to relax the bonds of authority, rather than bind them faster." That spirit required the magistrate to retreat from or resist entering three crucial areas: education, religion, and the economy. In all his political writings, these three concerns constantly intermingle and the perspective is the same. Happiness and progress require in all three "that the advantage we derive from them will be more effectually secured when they are conducted by individuals, than by the state" (*Writings*, 311, 305). Priestley makes specific the terms of the liberal revolution wherein much of human life, that hitherto had been seen as part of the single public social realm presided over by church, paternalistic officials, or moral economic norms, would be split off into different social spheres, there to be the concern primarily of autonomous self-regarding individuals.

Priestley was an outspoken foe of public education. His insistence that the state play no role in such matters repudiated both past practice, in which education was the particular concern of the state church, and the contemporary arguments put forth by his reformer friends Paine and Wollstonecraft, who were calling for active state involvement in a system of public education. Priestley's opposition was grounded in his doctrinal individualism. State education would hinder the flourishing of truth, variety, and improvement. It would indoctrinate whatever authorities take as truth or whatever serves their interest. It would restrict individual inquiry and development. The pursuit of truth is carried on by individuals who "dread nothing so much as being confined and cramped by the unseasonable hand of power." To put education "into the hands of the civil magistrate" would be to stifle forever freedom of inquiry; it "would be like fixing the dress of a child, and forbidding its clothes ever to be made wider or larger." That is much more than liberals intended the state to do, which is simply "the avenging of injuries, or redressing of private wrongs" (*Writings*, 308, 305).

Priestley develops even more self-consciously than Locke the theory of the liberal state as neutral umpire. The state has no longer any positive role to educate, nurture, or provide moral standards. It is neither concerned with the classical ideal of illuminating the good life or the republican ideal of providing an arena for fulfillment through public service and citizenship. The state, for liberal theory, has specific, limited, and negative functions which are implicit in its contractual origins. Its sole purpose is the protection of individuals and their rights. It is simply an agent performing the useful but limited service of keeping order and protecting individuals from harm. Nowhere is this liberal theory of the state better articulated than in Priestley's writings on the relationship of religion and the state. As in Locke before him, the most vivid articulation of the liberal theory of state is found in an explanation of why it must be separated from the church.

The state, according to Priestley, deals with "things that relate to this life," while the church deals with "those that relate to the life to come." The liberal state, then, is restricted to a specific purpose. It does no more nor less than provide a "secure and comfortable enjoyment of this life, by preventing one man from injuring another in his person or property." The magistrate has no concern with opinions or beliefs. His sole duty "is to preserve the peace of society." The state punishes only "if I break the peace of society, if I injure my neighbor, in his person, property or good name," not if I believe in different creeds. "How," Priestley asks, "is any person injured by my holding religious opinions which he disapproves of?" If the answer is that such opinions endanger the salvation of others, it is still inappropriate for the state to interfere, for its "business is with the things of this life only." Like Jefferson denying the liberal state the right to regulate religion because another's belief "neither breaks my leg nor steals from my purse," Priestley excludes the state from religious life as a use of civil power beyond the limited terms of the state's contractual base. For early liberals like Locke, Jefferson, and Priestley, the separation of church and state becomes the crucial defining feature of liberal politics. The realm of public power has been dramatically reduced. What right does the state or anyone else have to involve itself in my religion? Priestley asks Burke in his letter of 1791. "Does my conduct in this respect injure them? What, then, has the state, or my neighbors, to do in this business, any more than with my food or my medicine?" (*Writings*, 228; *Considerations on Church Authority*, 4; *Conduct*, 6; *Familiar Letters*, 4:18; *Letters to Burke*, 55).[2]

What Priestley, the scientist and liberal theorist, has set out to do is to demystify the state. Much more dramatically than Locke, Priestley the scientist seeks to free the state and political obligation from their conventional link to arcane religious sources, inscrutable and beyond human understanding. Deference to political power is severed from the timeless realm of hereditary and religious mystery. The state is rendered a simple and useful artifact, created by ordinary, self-seeking men whose rational common sense prompts them to consent voluntarily to be governed. Priestley exults in just the scientific demystification that Burke lamented. For Priestley, the state was indeed Burke's dreaded "nothing better than a partnership agreement in a trade of pepper and coffee, calico, or tobacco, or some other such low concern." It was, indeed, "taken up for a little temporary interest, and to be dissolved by the fancy of the parties." It was not to be looked on with "other reverence," as Burke insisted. Princes and magistrates, according to Priestley, are not shrouded in mysterious majesty. There are no rational grounds for "the superstitious respect for kings and the spirit of chivalry." Titles such as "most sacred or most excellent majesty" inaccurately describe mere "magistrates appointed and paid for the conservation of order." Political power is obeyed, not as Burke suggested from "generous loyalty," "proud submission," or "dignified obedience," but for the very mundane reason that "the good order of society requires it." Priestley will himself obey magistrates "only as the confidential servants of the nation, and the administrators of the laws" (Burke 1968, 194, 169; Priestley, *Writings*, 247, 255, 248).

Like the principles of the natural universe, these laws and the institutions of civil society are simple and easily comprehended. They require no complex, mysterious wisdom of specialized priests of statescraft, schooled by ageless traditions in the impenetrable secrets of politics. Stripped of metaphysical and religious mystery, stripped of aristocratic complexity, the state for Priestley would be streamlined and simple. It would merely administer justice and preserve the peace. Such a government, "being thus simple in its objects, will be unspeakably less expensive than it is at present," and thus the industrious would cease being taxed "the most unreasonable sum that we now pay for the single article of government" (*Writings*, 253).

Priestley constantly answers the Burkean defense of baroque mystery with this theme of simplicity and austerity. "You are told that matters of the state and church are of great mystery, into which you shouldn't delve," Priestley informs the dissenters of Birmingham in 1792. But such claims deceive, and he replies with a vintage articulation of the cult of simplicity.

> These things are not, in reality, of so difficult a nature as many things in your common trades and businesses. The most important questions [of state and church] are the plainest things in the world and require nothing but a common understanding, honestly applied, to comprehend them (*Familiar Letters*, 4:22).

Like Burke, who saw the age of chivalry replaced by the age of economists and calculators, Priestley also realized that European civilization was at a historic turning point. While this enraged Burke, it thrilled Priestley. "The whole of the Gothic feudal system," Priestley wrote in 1792, "embracing matters both of a civil and ecclesiastical nature, is beginning to shake to its foundation" (*An Appeal to the Public*, p. xiv). "The system which had stood for ages," he told his congregation with great joy on another occasion, was vanishing "like an enchanted castle in a romance" (*Sermons*, 101).

Burke and Priestley bear witness to the emergence of the secular liberal state, the partnership agreement to protect the trade in calico or tobacco, or, as it has been caricatured, the state as joint stock company. Obvious to both of them was how decisive the separation of church and state was in this demystification of the state. Burke, the secular man of affairs, repudiated the age of calculators and pleaded that "religion is the basis of civil society, and the source of all good and comfort" (Burke 1968, 186–87). Priestley, the scientist and Christian clergyman, replied with quintessential liberalism. The age of chivalry was indeed dead.

> Is there no good, or comfort in anything but religion, or what flows from it? Will religion feed or clothe us; or is there no comfort in food or clothing? Is it not possible to make many wholesome laws to prevent men from injuring one another, and is it not possible to execute those laws, so as to preserve the peace of society, which I conceive to be the proper end of civil government, without calling in the aid of religion? (*Letters to Burke*, 86).

Priestley's radical contemporaries shared his assumption that a new era was dawning in world history. The new order of scientific reasoning and self-interest would push aside mystery and superstition. Through it all breathes the cult of science and the cult of simplicity. It breathes through Paine's pleas for cheap, plain government and his constant evocation of the simplicity of American institutions. "The science of government is beginning to be better understood," Paine wrote in 1792. He was certain that "the age of faith and political superstition, and of craft, and mystery is passing away" (Conway 1895, 3:93). In religion, as in politics, he wrote in his *Age of Reason*, "perfection consists in simplicity" (ibid., 4:28). Godwin was another apostle of simplicity who in 1793 informs the reader of *Political Justice* that the recent events in America and France persuaded him of the "desirableness of a government in the utmost degree simple." There was "no mystery in government which uninitiated mortals must not presume to penetrate." The masses of mankind have been duped, Godwin claims, by those who talk of "the mysterious complicated nature of the social system" (Godwin 1976, 247, 552). Joel Barlow joins the attack on those who defend "government as the most complicated system ... without which it has been supposed impossible for men to be governed." He asks that his readers "conceive of the simplicity to which the business of government may be reduced, and to which it must be reduced, if we would have it answer the purposes of promoting happiness." Nature is simple and efficient; the old order is artificial, complicated, and ornate. "The essence of tyranny," wrote Barlow, "is to counteract the economy of nature, the essence of liberty is to promote it" (Barlow 1792, 172, 220).

This recurring tendency of Priestley's circle to link political arguments to conceptions of the natural world is central in understanding the radicalism of these English-speaking *philosophes*. Priestley's theory of the state is intimately connected to his scientific and theological materialism. He demystifies the state just as he demystifies the soul. In his essay on "The Materiality of the Soul" he notes "that there is not a simple idea of which the mind is possessed but what may be proved to have come to it from the bodily senses." The state, then, is what we touch, feel, and see. It is not any Burkean a priori ideal, transcending time and generations. It is the actions of real men, whose taxes impinge upon, and whose security protects physical beings. The state is a material entity, not a spiritual one. It acts on material beings through the careful distribution of rewards and punishments. "Governors," Priestley wrote, "will rule voluntary agents by means of rewards and punishments, and the governed, being voluntary agents, will be influenced by the apprehension of them." Government is for Priestley exactly what Burke denied it was — a thing of interest and calculation, not something spiritual and transcendental. There is no impenetrable and unknowable part of man, according to Priestley, "no immaterial soul." Just as the whole "substance of man is material," there being no spiritual part "capable of subsisting before and after its union to the body," there is no spiritual Burkean state that is more than the material state of calico and tobacco traders. The state is full of only material matter, that attracts and repels according to the gravitational laws of interest (*Writings*, 145–66, 97, 115; *History of Corruptions*, 5:480–81).

Despiritualizing the state renders it a mere machine, whose operations are not only simple and knowable but mechanical, predictable and manageable.

The state as machine was a recurring metaphor in the rhetoric of Priestley's circle. For Paine, the traditional British constitution was "like a complicated machine that never does right" (Conway 1895, 4:333). For Priestley, too, the old order was doomed in terms of the laws of mechanics. Writing of England's "Gothic" constitution, he notes that "the more complex any machine is ... the more liable it is to disorder" (*Writings*, 205). Robert Bage, the Birmingham dissenter and manufacturer, the friend of Priestley and an eminent novelist, was even blunter. In his novel *Barham Downs* he demanded that government become as straightforward, simple, and efficient as "a manufacture of buttons" (Bage 1784, 2:9).

If the state is but a mere machine, and if its inhabitants are mere matter, then politics is a fit subject for experimentation by the likes of Priestley and other citizen scientists. Priestley, indeed, discusses politics in these scientific terms. Since "we are so little capable of arguing a priori in matters of government," it is clear that "experiments can only determine how far this power of the legislature ought to expand." The magistrate should do little, in fact, "till a sufficient number of experiments have been made." Government is like everything else in this scientific world view. It stands "the fairest chance of being brought to perfection" if there "is opportunity of making the most experiments and trials." Priestley stood surveying the world and history as a political laboratory, where "hypothesis built upon arguments a priori are least of all tolerable, where observation and experience are the only safe guides" (*Writings*, 219, 307). The fruits of this scientific approach were facts of politics, overwhelming in their consistency.

> The old governments of Europe are arrived to a considerable degree of maturity; so that their several advantages and defects are become sufficiently conspicuous and the new governments in North America and especially those of France and Poland are so many new experiments of which political philosophers cannot fail to make the greatest use. (*Lectures on History and General Policy*, 39.)

It is, we must remember, Hartley whose influence lurks behind the scientific optimism of Priestley's materialism. Priestley's popularization of Hartley, whom he ranked with Newton, would make Hartley's ideas available after 1775 to dominate British social and psychological thought. As in the France of Holbach, Helvetius, and La Mettrie, advanced radical circles in Britain accepted the doctrine of the materiality of man and the parallel assumption of reform and improvement being a simple matter of changing the "associations" linked to external sensation. The materialist psychology, by eliminating the mind-body distinction, provided a scientific basis for the widespread belief that men's social and moral behavior could be changed by disciplining their bodies via pleasure and pain. The state literally becomes a laboratory where men and institutions can be perfected like machines by trial and error. For some, this is the task of the statesman. James Burgh, dissenting schoolmaster and most influential teacher of American and English radicals through his important *Political Disquisitions* (1775), wrote, "an able statesman can change the manner of the people at pleasure." It was clear to him "that by management the human species may be moulded into any conceivable shape" (Burgh 1775, 176–78). For others, this task was reserved for one special scientist, the medical doctor. La Mettrie himself had predicted in 1749, in his *L'Homme Machine*, that statesmen and judges would be replaced "by the most skillful physicians" (cited in Ignatief 1978). Indeed, the radical reformers in Priestley's circle were often medical doctors themselves, as had been Locke and Hartley. Five of Priestley's thirteen colleagues in the Lunar Society were medical doctors. Watt's first partner, Roebuck, was a physician who, like Kier and many others, turned to industry, especially chemical technology. The Manchester Literary and Philosophical Society, hotbed of dissenting radicals, was founded by Dr. Thomas Percival, a close friend of Priestley and the great hospital reformer. The great prison reformers Howard, Aikin, and Fothergill were medical doctors, dissenters, and friends of Priestley. Two important political reformers, John Jebb and Thomas Beddoes, were medical doctors. And, of course, one of the closest American friends of this circle was Dr. Benjamin Rush, himself a former student at the medical school of Edinburgh, where Hartleyan materialism was at the heart of the curriculum. The medical doctor, then, plays a crucial role in eighteenth-century social attitudes. Scientific student of the human body, he becomes the empirical wizard of the body politic, revealed now in all its simplicity as similarly a mere material mechanism of action and reaction.

The contemporary of Priestley who spoke most directly to this vision of the medical doctor was Priestley's close friend Thomas Percival. "Strangers to superstition and enthusiasm," doctors were, for Percival, the symbol of "the present enlightened period" in its crusade against "Gothic ignorance" and "barbarism." The doctor was the apostle of science, locked in combat with "the prejudices, the caprices, and the passions of the sick and their relatives." A crucial weapon in this struggle is the doctor's mastery of Hartleyan psychology, his "knowledge of the animal economy, and of the physical causes which regulate or disturb its movement and unites an intrinsic acquaintance with the laws of association" (Percival 1927, 73, 85, 87).

The doctor's role, for Percival, was not simply to cure the sick; he was also involved in social reform. His concern was the "moral improvement of the patients," which involved teaching them the principle of private property and the necessity of wills. The hospital had as its purpose disciplining and regulating the poor, much like Howard's prison and Wedgwood's factory. It was the antidote to the "immoderate passion," the "vicious indulgences," the "sloth, intemperance and irregular desires" which were the source of disease.

> It may be proved that a work hospital is an institution founded on the most benevolent principles, consonant to sound policy, and favorable to reformation and virtue. It provides relief for a painful and loathsome distemper, which contaminates in its progress the innocent as well as the guilty, and extends its baneful influence to future generations. It restores to virtue and to religion those votaries whom pleasure has seduced or villainy betrayed, and who now feel, by bad experience, that ruin, misery and disgrace are the wages of sin (Percival 1927, 86, 74, 122–23, 85).

Percival's hospital, locked and "removed from public inspection," was modeled on Howard's reformed prison. "The form best adapted (mutatis mutandis) to these essential paupers, appears to be that of the New Prison at Manchester, which is constructed on the well-known plan of Mr. Howard." Patients were kept in small separate rooms. They were to be isolated from society just as the new prison would end the easy access of friends and family to the prisoner. As Howard prescribed uniform clothes and washed walls for his prison, so Percival ordered the same for his hospital. Order and discipline would in both institutions conquer sloth and intemperance. Indeed, Percival's proposal for a Board of Health was offered as a form of "hospital police." These state officials would inspect and approve the accommodations of the poor. They would require that the houses of the poor be washed twice a year. They would pay a bounty of one or two shillings to anyone who reported "the appearance of fever in any poor family." These "hospital police" or "inspectors" would then give over the feverish poor to the physicians in Percival's hospital. When the fever was cured and the reformed returned to his family,

> a reward ... shall be given to the head of the family, after the cessation of the fever, on condition that they have faithfully observed the rules prescribed for cleanliness, ventilation and the prevention of infection, amongst their neighbors (Percival 1927, 86, 173, 180, 69, 168, 179).

Percival's hospital police would visit the poor, round up the sick, and send them to hospitals, there to be cured and reformed.

Priestley was not above social engineering himself. In 1787 he published

Account of a Society. Significantly enough, the scheme was intended primarily "for the use of John Wilkinson's (his brother-in-law) ironworks at Bradley." This Society was Priestley's alternative to the Poor Laws, which he saw as merely encouraging idleness. How, he asks, "to remove the Temptations to idleness and profligacy?" How to hold out "certain prospects of bettering his condition?" How to "enable him to do more for himself than he knows his parish will ever do for him?"

Priestley's answer involves giving the manufacturer the means with which to fix firmly in the minds of the poor the association between present self-denial and future pleasure. Priestley proposes that Wilkinson set up a "club" which involves forced deductions of a certain amount (4.5%) from the worker's pay each week. The manufacturer must teach "the men to provide for themselves." The manufacturers will create middle-class attitudes in the poor through their intervention. The poor will thus learn to be ambitious and competitive, "be enabled to vie with their neighbours in the acquisition of everything which is comfortable."

Priestley was not unaware of the tension between the disciplinary tone of his proposed "Society" and the emancipatory tone of his other writings. He acknowledges that this scheme involves "some restriction of liberty,..." with the worker "not allowed to spend his money in whatever way he thought best." Such restrictions are, however, according to Priestley, but invasions on brutes who are not yet self-directing individuals. Their intention is quite the opposite, in fact: to render the poor truly autonomous individuals who internalize the values of a truly free and therefore human person. These restrictions on freedom are thus necessary and just, for they see to it that "the idle and thoughtless may be made, in some measure, to do for themselves what the industrious and thoughtful are now compelled to do for them." What the Poor Laws do is quite the opposite, in fact: "keep them in a state of greater submission and dependence." His suggestion would set the poor on the path to middle-class freedom, for "they would be much more within the influence of honourable ambition than they are in at present" (*Account of a Society*, 3, 4, 7, 11, 13).

Priestley's utilization and popularization of Hartley helped shape the world view of an entire generation of dissenting factory masters, scientists, and reformers. Human beings and social institutions were, like the human body, material contrivances whose operations were knowable and manageable. Another figure in this period who faithfully followed Priestley's lead in this realm was his friend and follower, David Williams. Dissenting minister, schoolmaster, and friend of Franklin and Jefferson, Williams saw it as "the principal duty of man to transfer into social institutions, moral, civil, and political, the ideal he deduces from the natural world." Government should, then, be in the hands of "men of scientific minds." It was time, he wrote in 1782, "to discard the reveries of savages and apply the principles of science" to politics. Scientists governing would best understand the operations of the state, "the nature and construction of the machine" in which men live. Like so many of the others in Priestley's circle, Williams' mechanistic politics was fundamentally optimistic.

> As the development of a machine is owing to the prevalence of some constituent power or powers over others: so in a state, all inconveniences and injuries are to be ascribed to the want of sufficient counter-action and assistance in some of its parts, to balance the pressure of the others; and to assist in producing the ground effect (Williams 1810, 19, 188; idem 1782, 4–5, 74–75).

The men and women in Priestley's circle were the first generation to live with the machine and what they took to be its benign promise. The reality of the industrial revolution complemented the philosophical hegemony of Hartley. Mechanistic imagery is found everywhere. Wedgwood, pioneer in developing factory discipline with his bells and time clocks, wrote (in a letter to Thomas Bentley, October 1769) that he would "make machines of men as cannot err" (Wedgwood 1774, 82). Bentham's panopticon was conceived as "a machine for grinding rogues honest" (cited in Ignatief 1978, 69). Benjamin Rush, the American Hartleyan, wrote that "I consider it is possible to convert men into republican machines. This must be done, if we expect them to perform their parts properly, in the great machine of the government of the state" (Rush 1947, 92). Robert Owen, the industrialist reformer, who of all those schooled in psychological sensationalism would most have the opportunity to realize his utopian schemes in later years, once remarked that "the animate mechanisms" of New Lanark had been made as efficient as the "inanimate mechanisms" under his management (Owen 1813, 95). In a lighter mood, but in a most revealing one, Wedgwood reminds us how instinctive the use of mechanical imagery was among Priestley's friends. He wrote in 1782 to cheer up their fellow Lunatic James Watt, who was tired from overwork at his ironworks. Hartley, thanks to Priestley, was everywhere.

> Your mind, my friend, is too active, too powerful for your body and harasses it beyond its bearing. If this was the case with any other machine under your direction ... you would soon find a remedy. For the present permit me to advise a more ample use of the oil of delegation through your whole machinery. Seriously, I shall conclude in saying to

> you what Dr. Fothergill desired me to say to Brindley. "Spare your machine a little or like others under your direction, it will wear out the sooner by hard and constant usage" (Wedgwood 1915, 295).

There are two dimensions to Priestley's liberalism, which his connection to this radical use of Hartleyan psychology makes clear. There is a commitment to natural rights liberalism on the one hand and to utilitarian liberalism on the other. Priestley is a bridge between the two variants of liberalism. They are not incompatible for him because they both serve his overriding concern, reform, which is more pressing a passion even than freedom. One part of Priestley's liberalism preaches liberation, the freeing of all individuals and their rights from existing restraints — tyranny, the state, priests, and superstition. Another part of Priestley's liberalism subjects individuals to new discipline and new forms of authority: factories, jails, schools, hospitals run by scientific minds, scientific legislators, who teach, order, and manage men to become industrious and hard-working. This latter aspect of Priestley's outlook explains why he and his circle, men like Howard, Percival, and Wedgwood, were such passionate reformers. They were not simply interested in sweeping away ancient and feudal barriers that hindered a free and good society. They were also convinced that one could then move to a positive stage of managing, engineering, and creating a good and happy life that would not emerge in and of itself by merely getting rid of priests, tyrants, and superstition. Science and materialism cut both ways. It undermined the old order, liberating and freeing man from timeless domination and mystery. It also promised a new day when scientific leadership could produce great happiness for great numbers by manipulating men and their motions, even if achieving such happiness involved the sacrifice of freedom.

For all its commitment to emancipation and liberation, its zeal to free men from political, spiritual, and economic restraint, there is, then, in the radicalism of Priestley's circle an ominous specter of discipline, regimen, and authority. While this radicalism preached independence, freedom, and autonomy in polity and market, it preached order, routine, and subordination in factory, school, poorhouse, and prison. Eighteenth-century radicalism's liberating aspect was bent on toppling the aristocracy, its repressive aspect on improving the poor. Those in society who had not internalized self-reliance and self-discipline had to be bridled and tamed, taught to be methodical, clean, prompt, frugal, and industrious. Teaching and preaching middle-class values was one way to do this; another was through institutions over which these radicals presided — factories, schools, penitentiaries, hospitals. Here the moral fiber required for running the race of life could be acquired from time clocks, bells, and whitewashed walls, since the ideology's psychology assumed a malleable human nature ultimately shaped by association and environmental influence. The need for discipline would be transitory. It was not permanent subordination but a temporary restriction of freedom that Priestley's circle sought through their new institutions. This "temporary dictatorship of the middle class" would end when these new or reformed institutions had rendered everyone self-reliant and industrious. There was no doubt in Priestley's mind that such a day would dawn. In both its liberating and repressive aspects, scientific liberalism was indeed unabashedly optimistic.

Much more than simply the discoverer of oxygen, or one of Jefferson's "few lives precious to mankind," Priestley was the quintessential English *philosophe*, the reformer-scientist lifting the weight of mystery and tradition from both the universe and the polity. But there was also, as we have seen, an ominous side to Priestley's scientifically based liberalism. We began with Priestley's own linking of science and liberty, the former needing the latter to flourish. Yet, a science born in and requiring freedom from government and clerical intervention also became, for Priestley and his circle, an instrument curtailing the freedom of some. To be sure, even this was done in the name of liberal reform. It had to be, for it was science, more than anything else, that would allow men to see, as Priestley put it, "the morning opening upon us; and we cannot doubt but that, the light will increase, and extend itself more and more unto the perfect day" (*History of Corruptions*, 4).

Notes

1. The Priestley-Jefferson connection is a close one. On numerous occasions, Jefferson articulated the debt that his own educational, theological, and political attitudes owed to Priestley's ideas and writings. He read Priestley's *History of Corruptions* "over and over again" and it became the basis for his own peculiar blend of anticlericalism and deism. Through Priestley, Jefferson came to philosophical materialism, the associationist theories of David Hartley, and notions of the materiality of the soul. Jefferson's Millenarian vision of science and scientists is, at least in part, derived from Priestley's writings. The latter's writings on education were influential in Jefferson's planning of the University of Virginia, and Priestley's writings on the disestablishment of the Anglican church and the separation of church and state were used in 1786 by Madison and Jefferson in planning the statute for establishing religious freedom in Virginia. But Priestley and Jefferson shared more than just specifics. They shared an entire world view, an unabashed appreciation of modernity. Writing to Priestley in January 1800, Jefferson continued their running discussion on education. He agreed with Priestley that the study of Greek and Latin was not essential for a modern education. Jefferson then turned the letter to broader concerns, in a passionate defense of modernity against antiquity. "The Gothic idea

that we are to look backwards instead of forwards for the improvement of the human mind, and to recur to the annals of our ancestors for what is most perfect in government, in religion, and in learning, is worthy of those bigots in religion and government, by whom it has been recommended, and whose purpose it would answer. But it is not an idea which this country will endure." These were Priestley's views, too. He had written earlier, "Those times of revived antiquity have had their use and are now no more.... Their maxims of life will not suit the world as it is at present."

For documentation of Jefferson's debt to Priestley, see Ford 1905, 9:95, 102, 380, 404; 10:69. See also Bergh 1907, 10:228; 13:352; 14:200; 15:232. See also Koch 1964, 24, 27, 34; Boorstin 1960, 17–19, 113–19, 159–62; Bonwick 1977, 285; Han 1954, 404–26. For advice on commercial policy given Jefferson by Priestley and his disciple Thomas Cooper, see McCoy 1980, 176–77, 215–16, 246–47.

2. Like Locke, Priestly argues that the magistrate may interfere with religious practice if those practices themselves threaten life or rights (*Writings*, 230). Priestly goes beyond Locke, however, in his extension of toleration to Catholicism. For the Jefferson quote see his *Notes in Virginia*, Query 17.

References

I. Works of Priestley cited (referred to by abbreviated title):

Account of a Society = An Account of a Society for Encouraging the Industrious Poor to Which are Prefixed Some Considerations on the State of the Poor in General. Birmingham: 1787.

An Appeal to the Public.

Catechism = A Catechism for Children and Young Persons. London: 1787.

Conduct = Conduct to be Observed by Dissenters in Order to Procure the Repeal of the Corporatism and Test Acts. Birmingham: 1789.

Considerations on Church Authority. London: 1769.

Essay on Government = An Essay on the First Principles of Government. London: 1791.

Examination = An Examination of Dr. Reid's Inquiry into the Human Mind on the Principle of Common Sense. London: 1774.

Experiments on Air = Experiments and Observations of Different Kinds of Air. London: 1774.

Familiar Letters = Familiar Letters Addressed to the Inhabitants of the Town of Birmingham.

History of Corruptions = An History of the Corruptions of Christianity, in: The Theological and Miscellaneous Works of Joseph Priestley, ed. J. T. Rutt. London: 1819–32, vol. 5.

History of Electricity = The History and Present State of Electricity. London: 1769.

Lectures on History and General Policy. London: 1788.

Letters to Burke = *Letters to Rt. Honourable Edmund Burke*. Birmingham: 1791.

Letter to Pitt = *Letter to the Rt. Honourable William Pitt on the Subject of Toleration and Church Establishment*. London: 1787.

Memoirs = *Memoirs of Dr. Joseph Priestley*. London: 1806.

Proper Objects of Education = *Proper Objects of Education in the Present State of the World*. London: 1791.

Reflections = *Reflections on the Present State of Free Inquiry in This Country* (1785), reprinted in *Sermons*.

Sermons. London: 1830.

View = *A View of the Principles and Conduct of the Protestant Dissenters with Respect to the Civil and Ecclessiastical Constitution of England*. London: 1769.

Writings = *Priestley's Writings on Philosophy, Science and Politics*, ed. J. Passmore. New York: 1965.

II. Other works

Ashley Smith, J. W., 1954. *The Birth of Modern Education*. London.

Bage, Robert, 1784. *Barham Downs*. London.

Barbauld, Ann and John Aiken, 1792–96. *Evenings at Home; or, the Juvenile Budget Opened*, 6 vols. Philadelphia.

Barlow, Joel, 1792. *A Letter to the National Convention of France on Defects of the Constitution of 1791*. London.

Bergh, A. E., 1907. *The Writings of Thomas Jefferson*. Washington, D.C..

Bernal, J. D., 1965. *Science in History*. London.

Bonwick, C., 1977. *English Radicals and the American Revolution*. Chapel Hill, N.C..

Boorstin, D. J., 1960. *The Lost World of Thomas Jefferson*. Boston.

Burgh, James, 1775. *Political Disquisitions*. London.

Burke, Edmund, 1968. *Reflections on the Revolution in France*, ed. C. C. O'Brien. Harmondsworth.

—, 1791–92. "Speeches on Fox's Motion for Repeal of Certain Penal Statutes Respecting Religious Opinions," *Parliamentary History* 29.

Chaloner, W. H., 1958. "Dr. Joseph Priestley, John Wilkinson and the French Revolution 1789–1802," *Transactions of the Royal Historical Society* (5th Series) 8.

Conway, M. D., ed., 1895. *The Writings of Tom Paine*. New York.

Cooper, Thomas, 1812. *Introductory Lecture of T. Cooper, Professor of Chemistry at Carlisle College*. Carlisle.

D'Elia, D. J., 1970. "Benjamin Rush, David Hartley and the Revolutionary Uses of Psychology," *Proceedings of the American Philosophical Society* 114(2).

Dobree, B., ed., 1935. *The Letters of King George III*. London.

Ford, P. L., 1905. *The Works of Thomas Jefferson*. New York.

Fruchtman, J., Jr., 1983. *The Apocalyptic Politics of Richard Price and Joseph Priestley: A Study in Late Eighteenth Century English Millennialism*. Philadelphia.

Gibbs, F. W., 1967. *Joseph Priestley: Revolutions of the Eighteenth Century*. New York.

Godwin, William, 1976. *Enquiry Concerning Political Justice*, ed. I. Kramnick. Harmondsworth.

Han, N., 1954. "Franklin, Jefferson, and the English Radicals at the End of the Eighteenth Century," *Proceedings of the American Philosophical Society* 98:406–26.

Hartley, David, 1749. *Observations on Man, His Frame, His Duty, and His Expectations*. London.

Holt, A. D., 1931. *A Life of Priestley*. London.

Ignatief, M., 1978. *A Just Measure of Pain*. New York.

Jefferson, Thomas, 1972. *Notes on the State of Virginia*. Norton.

Johnson, Samuel, 1775. *Taxation No Tyranny*. London.

Koch, A., 1964. *The Philosophy of Thomas Jefferson*. Chicago.

Kramnick, I., 1980. "Children's Literature and Bourgeois Ideology: Observations on Culture and Industrial Capitalism in the Later Eighteenth Century," in: *Culture and Politics: From Puritanism to the Enlightenment*, ed. P. Zagorin. Berkeley.

—, 1982. "Republican Revisionism Revisited," *American Historical Review* 87.

McCoy, Drew, 1980. *The Elusive Republic*. Chapel Hill, N.C.

McKendrick, N., 1961. "Josiah Wedgwood and Factory Discipline," *Historical Journal* VI(1).

McLachlan, H., 1931. *English Education under the Test Acts*. Manchester.

—, 1943. *Warrington Academy, Its History and Influence*. Manchester.

Malone, D., 1961. *The Public Life of Thomas Cooper*. Columbia, S.C.

Marshall, L. S., 1946. *The Development of Public Opinion in Manchester 1780–1820*. Syracuse.

Mill, John Stuart, 1924. *Autobiography*. Oxford.

Owen, Robert, 1813. *A New View of Society*. London.

Parker, I., 1914. *Dissenting Academies in England*. Cambridge.

Percival, Thomas, 1803. *Medical Ethics*. London.

Robinson, E., 1953. "The Derby Philosophical Society," *Annals of Science* 9.
—, 1956. "The English Philosophes and the French Revolution," *History Today* VI(2).
Rush, Benjamin, 1947. *The Selected Writings of Benjamin Rush,* ed. D. Runes. New York.
Rutt, J. T., ed., 1831. *The Life and Correspondence of Joseph Priestley.* London.
Schofield, R. E., 1963. *The Lunar Society of Birmingham.* Oxford.
Taylor, B., 1954. *Joseph Priestley: The Man of Science.* London.
Taylor, J., 1943. "The Sheffield Constitutional Society," *Transactions of the Hunter Archaeological Society* 5.
Wedgwood, Josiah, 1974. *The Selected Letters of Josiah Wedgwood,* ed. A. Finer and G. Savage. London.
Wedgwood, Julia, 1915. *The Personal Life of Josiah Wedgwood.* London.
Willey, B., 1961. *The Eighteenth Century Background.* Boston.
Williams, David, 1782. *Lectures on Political Liberty.* London.
—, 1810. *Preparatory Studies for Political Reformers.* London.

Science, Politics, and the New Science of Politics
A Comment

JOSEPH MALI

The study of English and American radicalism in the eighteenth century has become the focus of a lively historical debate in the last two or three decades. The renewed appreciation of the radical Enlightenment owes its inception to the liberalism of the 1960s, which sought its intellectual origin in that era. We can but derive pleasure and hope from the fact that this search is still going on in our days of social and political Neoconservatism.

Isaac Kramnick has been one of the leading scholars in this field and his contribution to the history of ideas is as important as it is well known. In his illuminating paper he has acquainted us with Joseph Priestley, and in so doing has presented us with a typical case of the Enlightenment natural scientist turned social theorist. His main argument, that Priestley's scientific activity and philosophy radicalized not only his social and political ideas but also the social and political institutions of his time, was carried through with remarkable clarity. Priestley himself emerges from this presentation as a saint of modern science, almost immune to any critical commentary. And here is my problem: I am not an expert on Joseph Priestley. Neither am I an expert on the history of science, or even on Anglo-American history. All I can do is add a few remarks about Priestley's methodology. Talk of methodology, I admit, is the last refuge of the ignorant.

Peter Gay, the modern historian of the Enlightenment, has argued convincingly that the Enlightenment should be seen not as the Age of Reason but rather as the Age of Criticism. In Joseph Priestley we have a scientist who was totally committed to the task of criticism. His critical activity in the

E. Ullmann-Margalit (ed.) The Scientific Enterprise, 33–41.

natural and social sciences can be summarized in two words: ALWAYS DEMYSTIFY! This was his philosophical credo and political battle cry. His demystification of the natural world and the polity was radical both in its theoretical assumptions and in its practical applications.

Kramnick has rightly emphasized that Priestley's radicalism consisted in his attempt to reduce all nonmaterial essences in the natural human spheres to pure mechanistic-materialistic components. There can be no doubt as to the adequacy and success of this method in the study of nature. Much more problematic, however, was his attempt to demystify the polity, to radicalize our conceptions of man and society. Karl Marx remarked once that "to be radical is to grasp matters at their root. But for man the root is man himself." This means that the crucial aspect in any science of man is what image-of-man can be laid bare. In Priestley's case we have seen that his image-of-man was thoroughly that of man-machine, based on Lockean psychology and modified by Hartley's associationism. This mechanistic-materialistic image of a man determined a similar model of society and a science of politics. My commentary will center on Priestley's science of politics.

The problem, simply put, is this: How are we to evaluate Priestley's attempt to construe a science of politics modelled on the same mechanistic-materialistic principles of his science of nature? While we may all agree on and applaud its practical success in "demystifying the polity," we may raise some doubts about its theoretical foundations and claims. We must ask whether the basic ideas and methods of his new science of nature, applied to his radical politics, merged into a truly new and radical science of politics; or whether, when judged by the standards of countertheories of politics in his time, his enterprise marks the end of a tradition, rather than announcing a new one. My contention will be that Priestley's mechanistic imagery was indeed most innovative and useful in destroying old conservative myths. But, one must ask, did he thereby create a new science of politics, that is, a better theory with which to understand and explain politics? My answer would be that he did not. I would argue that, because of his rigid, positivistic reductionism of all human phenomena and sciences to a mere physicalism, Priestley ultimately failed to understand and account for what was recognized by other theorists in his time as most unique to man and to the study of man, namely, culture. As such, Priestley's application of his scientific method to politics had more in common with the old naturalistic science of politics, practiced in the seventeenth and eighteenth centuries, than with the truly new, cultural science of politics, which emerged in the late eigteenth and early nineteenth century.[1]

Priestley's faith in rational perfection, social reform, and political progress through science, and only science, is akin in its ideological thrust to the "cult of science" and naive programs of the socialist utopians of the early nineteenth century. It seems to me that truly radical theorists of society took their inspiration from the counter-Enlightenment tradition in the human sciences. I would argue that the modern science of politics, which emerged toward the end of the eighteenth century, owes more to those thinkers who actually rejected Priestleyan ideas and methods, thinkers who were indeed his political adversaries.

Priestley's analytical approach to things political threatened conservative organicists like Burke, since it entailed a radical critique of all political myths of origins and privileges. Burke's conservative reaction, as exemplified in his *Reflections on the Revolution in France* (1853), had to address itself not merely to new radical claims, but also to a new kind of political discourse. He had to counter a new, radical science of politics which used abstract, analytical, and critical arguments, thus proving inimical to his ideas and rhetoric about things political.

Kramnick, speaking for Priestley, contrasted Priestley's style of "simplicity and austerity" with Burke's "baroque mystery." Burke, indeed, repudiated the angular rationalism of the eighteenth century and objected to abstract analysis in politics, because his aim was the opposite: to reassert and sanction a mystical dimension for society. Burke condemned all philosophical speculations on political matters because, as he said, "The science of constructing a commonwealth and of renovating or reforming it, is — like every other experimental science — not to be taught a priori." Like Priestley then, Burke too perceived the new science of politics to be an experimental science. He thought, however, that his own inquiry into things political, with its emphasis on the irrational forces that motivate people, was more faithful to political reality and therefore formed a better science of politics.

We see then, that Priestley and Burke were as much divided in their views about the study of politics, and what constitutes it as a truly experimental science, as they were divided in their actual political views. It seems to me that, both generally speaking and in reference to Kramnick's paper, the conservative arguments, precisely because of their conservative or reactionary themes, have not been appreciated for their innovative methodology. The fact that the "conservative" voices often argued against science does not necessarily mean that their methods of inquiry and argument were equally antiscientific. In fact, as I would like to suggest, the romantic reaction did not only produce mystical beliefs; it also developed a keen cultural and historical sensibility, which proved to be crucial for our modern conceptions of the humanities and social sciences.

I would now like to elaborate on the differences between the two approaches to politics. In order to accentuate these differences my presentation will be rather schematic.

During the eighteenth century, the project of creating a science of politics acquired new vigor and dimensions. This was part of the general attempt of scientists and freethinkers to apply the Newtonian or experimental method to the human spheres. The problem was how to reduce the chaotic information about politics to systematic knowledge and objectivity in the assignment of causes. Hume's definition of politics as dealing with "men united in society and dependent on each other" determined its frame of reference. There were two basic approaches. One sought to draw lessons or maxims from historical evidence. History, in Bolingbroke's famous words, was no more than "a philosophy teaching by examples." The modern secular realm of political history, stripped of its theological meaning, thereby acquired a practical significance. The other approach, which became prominent through the works of Hobbes, Locke, and Hume, concentrated instead on the constant "principles" or "springs of actions" in human nature, those psychological propensities that seemed not to vary across time and place. Hume was cautious but confident that the task and his approach were feasible. It was necessary, he said, "to glean up our experiments in this science from a cautious observation of human life, and take them as they appear in company, in affairs, and in their pleasures" (Hume 1888, p. xxxiii). Hume believed that local experience was significant and useful only insofar as it attested to one universal and unchanging human nature:

> It is universally acknowledged, that there is a great uniformity among the actions of men, in all nations and ages, and that human nature remains still the same, in its principles and operations. Would you know the sentiments, inclinations, and course of life of the Greeks and Romans? Study well the temper and actions of the French and English: You cannot be mistaken in transferring to the former most of the observations which you have made with regard to the latter. Mankind are so much the same, in all times and places, that history informs us of nothing new or strange in this particular (Hume 1902, 83).

Since men are, and have always been, the same, there was no need or reason to enquire deeply into the various forms and norms of their unique political life. What mattered to Hume and his ilk was man's constant and uniform nature as it *manifests* itself in politics, and man's accidental and diverse temperament as it *expresses* itself in the political lore of national communities. The methodological postulate that the science of politics, as indeed all human sciences, must be modeled on the natural sciences was self-evident. That Priestley shared these basic themes and ideas is, I think, quite clear and need not be elucidated further here.

The French Revolution was, for these theorists, a perfect experimental model. They saw in it a case which was actually carried into practice, and could also be theoretically analyzed according to the very principles of rationality which they had espoused. Their political adversaries, however, considered this revolution an experimental case which proved exactly the opposite: that human actions are irrational, that organic and spontaneous forces — and not abstract analytical ideas — motivate man. The conclusion was that the science of politics must therefore account for these actions with new means, adequate to the challenge.

From around the middle of the eighteenth century onward, European scholars began to develop new ideas about the subject matter and scope of the science of politics. The encounter with the variety and diversity of human cultures in the New World, but also in their own nations, revealed to these scholars that their uniformistic conceptions of human nature were too schematic and limited. In view of the many different examples of human nature which confronted them, they realized that their adherence to the Newtonian mode of analysis and reduction to "first principles" resulted in the deduction of some universal human characteristics which were forced and useless. A science of politics which was confined to Western civilization and to its taxonomy of politics, on the basis of classical forms and norms, proved equally inadequate. Priestley is a case in point: his naive faith in the power of the synthetic-instrumental method obscured from him the fact that political inquiry was qualitatively different from scientific experiment. His political adversaries, most notably Burke, seem to have recognized this fact.

The failure of the human sciences to deal with the new cultural and anthropological phenomena was to become a major concern for thinkers like Voltaire and Montesquieu in France, Vico in Italy, Bodmer in Switzerland, and Herder in Germany. Ultimately, they saw that the cause of that failure lay in the inadequacy of their scientific model and naturalistic explanations. The humanities and social sciences were now recognized as essentially different because, as Vico phrased it, "the world of civil society has certainly been made by men, and its principles are therefore to be found within the modifications of our own human mind." Vico differentiated between "the world of Nature, which, since God made it, He alone knows," and "the world of nations, or civil world, which, since men had made it, men could come to know" (Vico 1968, par. 331). The knowledge of the world of nations differed from the knowledge of the world of nature not only in a degree of certitude, but in a kind of comprehension. What was required, according to Vico and similar-minded theorists, was a new science of politics, which would not be a strictly experimental science in search of general natural laws, but rather an interpretive science in search of some specific cultural meaning.

Montesquieu's concept of *Esprit*, Vico's concept of *sensus communis* and Herder's *Geist des Volkes* are akin to Burke's notion of "prejudices." They all signify a new awareness as to the cultural essence of the study of society. Or, in view of Marx's remark about being truly "radical" in the human sciences, we see that these theorists operate with a new image-of-man, which is counter-rational and therefore politically reactionary but, in its own way, radical. They perceive man, society, and the state as essentially mythopoeic, that is, as originating from and consisting in what Vico called the "three principal operations of the mind ... imagination, memory, and ingenuity."

Thus, the task of the science of politics, as of all human sciences, is not really a scientific one, but again to use Vico's term a "*new* scientific" one. It is a hermeneutical, not a mechanical task. This science does not ask whether a political view or action is right or wrong in absolute terms, but rather what it means.

Before I conclude I should like to illustrate the above point with the aid of two examples: the semantic differences between Priestley and Burke. This will help us to see the essential difference in their scientific approach to politics as well. Take, for example, their antagonistic attitudes toward myths. Priestley calls them "superstitions," and discards them as mere lies, cunning fabrications of the rulers devised in order to deceive the people, and so on. Burke calls them "prejudices," but for him this term has positive, even constructive connotations. Myths are the common habitual attitudes, popular beliefs, and traditions which are indispensable to society, as well as to those who wish to understand society's ways. Burke's position, here and throughout his lifelong battle against Priestley and the rationalist critics of traditional authority, hinged on his contrary image of man:

> In this enlightened age I am bold enough to confess that we are generally men of untaught feelings, that, instead of casting away all our old prejudices, we cherish them.... We are afraid to put men to live and trade each on his private stock of reason, because we suspect that this stock in each man is small, and that the individuals would do better to avail themselves of the general bank and capital of nations and ages.

Burke maintains that political leaders and theorists must pay greater attention, as well as homage, to these prejudices. As against the iconoclastic zeal of the radical parties in France and elsewhere, he praises the conservative effort of "our men of speculation" who,

> instead of exploding general prejudices, employ their sagacity to discover the latent wisdom which prevails in them. If they find what they seek, and they seldom fail, they think it more wise to continue the prejudice, with the reason involved, than to cast away the coat of prejudice and to leave nothing but the naked reason; because prejudice, with its reason, has a motive to give action to that reason and an affection which will give it permanence (Burke 1853, 2:359).

Or, if we take the term "common sense," which both Priestley and Burke use in more or less the same meaning, but with contrasting connotations, we see the same pattern. Priestley's polemics against the philosophers of common-sense of his time — Thomas Reid, James Beattie, and James Oswald — were motivated by his strict rationalist conviction that the associationist principles of Locke and Hartley are the only valid road to truth. The appeal to common sense (or, in Priestley's derogatory phrase, the mere "sense of truth") seemed to him a violation of all the rules of scientific inquiry. It entailed the dangerous claim that sensual perception is scientifically valid insofar as it is common, and thus that it can be an alternative source of truth beyond reason. The extinction of reason, he feared, was imminent if prompt measures were not taken:

> Considering the very late origin of this new empire of common sense, its conquests, it must be confessed, have been very rapid; and as it has subdued all the regions of metaphysics, morals, and theology in the space of ten years, it may be computed that, with this addition of strength, it may, in ten years more, complete the reduction of all the seven sciences; when the whole business of *thinking* will be in a manner over, and we shall have nothing to do but to *see* and *believe* (Priestley 1774, 201–202).

Priestley was alien, or even hostile, to the notion that in the sphere of human affairs, at least, we seek a different kind of "truth," namely an idea, common and practical, which people hold "to be true" and according to which they regulate their lives. He could not possibly understand, let alone accept, the methodological insight that this "truth," or verisimilitude, could be found by means of inquiry which are more "sensitive" to its particular aesthetic and popular dimensions.

For Burke, common sense was an aesthetic image of life, spontaneous and organic to each community. It denotes those archaic images and symbols which are intuitively shared by members of the community, those which define and unite them in the common sensibilities of their tradition. He firmly believed that "truths" in politics, history, morality and the like can be found only through and in "common sense."

Both Priestley and Burke regard the search for common sense in politics as crucial for their science of politics; but while Priestley seeks to abrogate its validity in philosophical discourse and political activity, Burke hails its philosophical authenticity and political value. Surely, Priestley's search has more radical ambitions and impact, but the question as to whether his rigorous search is also more radical in its insights and implications for the human sciences is still open. It seems, paradoxically enough, that with regard to method, at least, the reactionary Burke is slightly more innovative and instructive than the progressive Priestley.

The new science of politics which emerged in the nineteenth century steered a middle way between these two extreme positions — and away from them. Both Priestley, the radical realist, and Burke, the conservative idealist, were superseded by thinkers who had much better historical perspectives and more refined psychological and anthropological conceptions about men as "political animals." "The passion for realism" in politics and elsewhere, which was so characteristic of the nineteenth century, cannot be elaborated on here. It can only be hinted at, by reference to a contemporary of Priestley and Burke, Edward Gibbon, who stood as it were between the two. Gibbon, the philosophic historian, was, like Priestley, most hostile to all those prejudices and superstitions which he found in Burke's writings. At the same time, as a conservative politician, he was aware that these "fleeting shadows of metaphysics," destructive as they were to human reason, were nonetheless indispensable and quite constructive to political thought and life. Committed as he was to the cause and ideals of the Enlightenment, Gibbon saw, through his investigations into history, that men, in pursuing their political ideals, have never thought or acted according to the rational ideas of "the truth," but only according to their credulous and fabulous ideas of "the true." In the chaos of time these illusions gave their lives and history some necessary order, meaning, and purpose. For Gibbon, the study of politics, much like its making, had become again what it had been in classical times — a practical rather than a theoretical science. It must be an empirical search for what was, is, or can be "the true" for man, a search which will explain, but by all means retain, that which is most essential to any body politic: its authority. Upon reading Burke's *Reflections on the Revolution in France* Gibbon remarked: "poor Burke is the most eloquent and rational madman that I ever knew": "madman", because Burke really believed in his superstitions, but an "eloquent and rational" one, because he knew why he was holding and preaching for them. For all these reasons Gibbon could say: "I admire his eloquence, I approve his politics, I adore his chivalry, I can even forgive his superstition" (Norton 1956, 3:229, 216). And so, perhaps, with all due respect for Priestley and for the Enlightenment ideals of rationality, science and progress, can we. Our modern political experience and views have revealed to us the truth of Burke's pessimistic yet sublimative assertion that in politics, as in any other collective effort to improve the human condition, "human beings build their cultures, nervously loquacious, upon the edge of an abyss."

Notes

1. For a general philosophical account of this new science of politics, see Wolin 1960, chap. 9 and 10; for an excellent historical study of the emergence of new theories in Britain, see Collini, Winch, and Burrow 1983.

References

Burke, Edmund, 1853. *Reflections on the Revolution in France*, in: *The Works of Edmund Burke*, London: Bohn's Standard Library, vol. 2.

Collini, S., Winch, D., and Burrow, J., 1983. *That Noble Science of Politics*. Cambridge: Cambridge University Press.

Hume, David, 1888. *A Treatise of Human Nature*, ed. L. A. Selby-Bigge, Oxford: Oxford University Press.

—, 1902. *An Enquiry Concerning Human Understandings*, ed. L. A. Selby-Bigge, Oxford: Oxford University Press.

Norton, J. E., ed., 1956. *The Letters of Edward Gibbon*. London: Cassell and Co.

Priestley, Joseph, 1774. *An Examination of Dr. Reid's Enquiry into the Human Mind on the Principles of Common Sense*. London.

Vico, Giambattista, 1968. *The New Science*, trans. T. G. Bergin and M. H. Fisch, Ithaca, N.Y.: Cornell University Press.

Wolin, S., 1960. *Politics and Vision, Continuity and Innovation in Western Political Thought*. London: George Allen & Unwin.

Critical Reactions to the Occult Sciences During the Renaissance

BRIAN VICKERS

> Men seldom seek a high degree of proof
> for what they already believe to be true.
>
> Thomas 1973, 657

The revival of the occult sciences — astrology, alchemy, numerology and natural magic — during the Renaissance used to be an embarrassment to historians. Some preferred to address themselves to the Scientific Revolution and simply ignore the occult leanings of Ficino, Pico della Mirandola, Giordano Bruno, John Dee, Robert Fludd, and others. If one scholar changed this conspiracy of silence it was Lynn Thorndike, in his *History of Magic and Experimental Science*, published in eight volumes between 1923 and 1958. This vast survey — as Eugenio Garin described it, "a catalogue and index rather than a history" (Garin 1983, 133, n. 8)[1] — is much more reliable on magic than on science; indeed, as his work progressed, Thorndike came to identify rather with the occult tradition, and could barely bring himself to discuss its opponents with either objectivity or fairness. Thorndike's unrivaled knowledge of manuscript materials and his indefatigable researches certainly laid a new scholarly foundation for studies in the occult sciences; and his work, together with the tradition established by Eugenio Garin in Florence (including Paolo Rossi and Cesare Vasoli), can be seen as both an inspiration and a reference point for the sudden popularity of magic in the English-

E. Ullmann-Margalit (ed.) The Scientific Enterprise, 43–92.

speaking world in the 1960s and 1970s. One other source of influence was the Warburg Institute, whose library and teaching staff reflected the full range of Aby Warburg's interests, including astrology and magic. D. P. Walker, whose *Spiritual and Demonic Magic from Ficino to Campanella* (1958) was a key work in this tradition, moved from the Department of French of University College, London, to the Warburg Institute in 1961, while Frances Yates had been a Research Fellow of the Institute since 1938. Also influential was the work of Walter Pagel, of the Wellcome Institute for the History of Medicine, with his studies of the Paracelsian tradition. Younger scholars influenced by these diverse groups include Paola Zambelli, P. M. Rattansi, and Charles Webster.

Anyone who cares about the total history of any historical period could only welcome this new interest in the occult tradition. Yet its proponents were sometimes carried away by the exciting novelty of their work and made over-large claims for the occult, elevating it to the status of a dominant or even defining element in Renaissance culture. Frances Yates, for instance, in an essay entitled "The Hermetic Tradition in Renaissance Science" (Yates 1967),[2] celebrated what she called "the Renaissance magus," with his "powers of *operating on the cosmos* through magic and through the numerical conjurations of cabala" (my italics). According to her, such "magi" as Ficino, Cornelius Agrippa, and John Dee provided "the chief stimulus of that new turning toward the world and *operating on the world* which, appearing first as Renaissance magic, was to turn into seventeenth-century science" (my italics) (Yates 1967, 255, 257, 272). In her work on the Rosicrucians, Yates (1972) came to see this "movement," too, as causal and operative in Renaissance science (see Vickers 1979). Throughout her work one looks in vain for any account of the other side of this argument, and in those who followed her a similar one-sidedness is apparent. The unwary reader is given the impression that magic somehow dominated Renaissance culture, and that to criticize it would be the mark of crass nineteenth-century positivism or the Whig concept of history as a triumphant procession of inventions and discoveries. The result has been a lopsided view of Renaissance culture. While everyone agrees that the occult sciences were finally discredited in the late seventeenth century — a change marvelously documented by Keith Thomas in *Religion and the Decline of Magic* (Thomas 1973) — there is no general agreement as to the causes of their eclipse. While I do not attempt to settle that debate here, I believed that more attention should be given to those Renaissance texts which produced an articulate critique of the occult, challenging its assumptions, methods, and goals. This largely overlooked movement, with a sustained flow of criticism between 1490 and 1690, from

Pico to Bayle, was an important element in the downgrading of the occult as an intellectually respectable scientific activity.

Given the vast scope of the European Renaissance, which extends as a coherent movement from Petrarch to Milton, say, and the great diversity of the occult sciences (a useful guide is Shumaker 1972), in order to keep this essay within manageable bounds I shall concentrate on one discipline — astrology — and on some of its relations to botany, medicine, and chemistry. And since the Renaissance critique of astrology is so detailed that it cannot be understood without a firm grasp of the astrological tradition, I shall first analyze that. My aim, however, is not to summarize astrology as a body of doctrine, but rather to draw attention to the mental processes that lie behind it, its modes of conceptualizing reality and "working" on it.

I

Despite some claims that the two disciplines were invariably confused, astronomy and astrology were often clearly separated in literature and in practice: astronomy studied the motions of the planets, astrology made deductions and, above all, predictions about human behavior from them. Whoever studies Ptolemy's *Syntaxis*, or other extant works of Greek astronomy, or reads the modern accounts of them given by Otto Neugebauer (1957), B. L. Van der Waerden (1965), Marshall Clagett (1957) and Geoffrey Lloyd (1970, 1973, 1979), is likely to be impressed by the level of accuracy attained by observers two thousand years before the invention of the telescope. The extent of their star maps and catalogs; their observation of planetary movements; their formulation of a cosmic system — in all these fields the Greeks, building on knowledge acquired from other cultures (notably Babylonian), made remarkable advances. Greek astronomy is observational and mathematical. Greek astrology is only partly observational, for on to the observational element — the movement of the planets, the course of the Sun through the zodiac — it superimposed a level of anthropomorphic symbolism, which became in turn the basis for an elaborate process of classification and correlation.[3] The crucial defining characteristic of astrology is this triple movement of attributing human qualities to the planets and zodiacal stars; then interposing various systems of distribution and correlation that yield greater possibilities of classification; and finally applying these data to the individual case. In the following brief exposition I shall survey first the ascription of human attributes, what one might call anthropomorphic or cultural categories (sex, temperament, place in a hierarchy), then the intermediate systems of division and classification.

The initial assumption of astrology is that the stars exert a dominating influence over mankind. In the Hellenistic period, as the great French scholar Bouché-Leclercq wrote, "for the adepts of the mystical philosophy derived from Pythagoras and Plato there was no need to demonstrate this proposition"[4] since the stars were conceived of as "Gods, and even gods who are creators of this world's organisms" (B-L, 73).[5] Few later exponents of astrology felt the need to prove this assumption; it was simply believed. The most important agents of influence were the seven planets — the Sun, the Moon, Saturn, Jupiter, Mars, Venus, and Mercury — and the twelve signs of the zodiac. Since the Greeks conceived their gods in anthropomorphic terms, the planets, which were named after the gods, also acquired human characteristics, both physical and psychological. To the Sun astrologers could attach all kinds of epithets: Titan, Hyperion, Phoebus and even Phaeton, which — as Bouché-Leclercq points out — "could give rise to mistakes, since it is the name of the planet Jupiter" (B-L, 90n.).[6] The Moon, celebrated as the cause of all change on Earth, was transformed by the Greeks from the masculine star of the Chaldeans to a feminine body, for reasons partly mythological, partly scientific (menstruation), and was connected with the humid element (B-L, 91–92). The anthropomorphic attitudes emerge most clearly with Saturn, the planet which takes longest to circle the Earth (some 29 years) and has a rather livid appearance: on both counts Saturn was pictured as old, serious and malign. Mythology and etymology associated him both with the Chronos, who mutilated his father, and with Χρόνος as time.

> Whence arose a series of associations of incoherent and divergent ideas, in which astrologers chose at will the elements composing the type of the powerful and redoubtable star, enthroned in the highest heaven, which presses irresistibly on celestial levers. (B-L, 94)[7]

The way in which this "swarm of analogies and permutations," "all of these incoherent ideas,"[8] make up "the astrological type of Saturn"[9] (B-L, 94, 97) shows the peculiar two-way anthropomorphism basic to astrology. That is, human qualities are first attributed to the stars, and then deduced from them.

Jupiter, in contrast to Saturn, was the most benevolent and beneficent of the stars; if his influence alone could dominate, the Earth would be paradise, one ancient astrologer said, and men would be immortal (B-L, 97). If Jupiter is "tempered" or well balanced, Mars, whose course, seen from the Earth, "looks like rapid leaps following upon short retrograde motions"[10] — was characterized as "impetuous" and "flamboyant." His reddish, blood-like color, and his association with Ares, the god of war, caused him to be described as a tyrant over the human race and the cause of disturbance in the whole of nature (B-L, 99). Venus, inevitably associated in mythology with Mars and her famous adultery, was feminine and humid, like the moon, but with added heat. She is both the morning and the evening stars, which were originally conceived as two beautiful boys loved by Aphrodite, who came to swallow them up (B-L, 99–100). To Mercury or Hermes were given the associations of lightness, instability, with a Protean ability to change in either direction, due to the apparent rapidity of his course and the influences he received from neighboring planets (B-L, 100–101).

The ascription of qualities to the planets derived in this way from mythology and anthropomorphic religion. It also derived from physics, yet the apparently neutral classifications of Aristotelian physics that are used by Ptolemy in his *Tetrabiblos* (I.4) — hot, cold, humid, dry — were also put to anthropomorphic use, a process which produced conflicts between physics and astrology. Saturn was deemed to have the greatest influence, even though he was at the highest point in the sky, being nearest to the fixed spheres where the gods resided. According to physics, then, Saturn should have been equated with fire, the element that, because of its inherent lightness, occupies the highest part of the world. But this did not suit the astrologers, who conceived of him as a cold and morose old man (B-L, 95). The relation between temperature and sexuality reappears with Venus, who is hot and humid, hence fertile. By a further extension of physical categories into an anthropomorphic frame, the astrologers classified heat and humidity as "active and generative" principles, cold and dryness as "passive and destructive" ones. It followed that the planets associated with the first category — Jupiter, Venus, the Sun, and the Moon — are beneficent; the rest — Mars, Saturn, and Mercury in part — are maleficent (*Tetrabiblos*, I.5; B-L, 101). Here one can note the tendency of astrology to use binary oppositions in forming categories, a habit which, together with the deterministic implications, caused its critics to describe it as Manichaean.

Anthropomorphic physics also determined the sexes of the planets. A predominance of humidity produced feminine planets, the Moon and Venus, while both physics and mythology ascribe masculinity to the Sun, Jupiter, Mars, and Saturn. Mythology made Mercury male, but astrological physics made him bisexual: he is said to be a mixture of the humid (female) and the dry (male), because, as Bouché-Leclercq puts it, the astrologers wanted to introduce symmetry into their classification (B-L, 101). In general the superior planets are masculine, the inferior feminine, and Mercury a kind of go-between (B-L, 102). But other categories were brought in to change the sex of the planets, such as their "phases" in relation to the Sun. Those that rise in the morning before the Sun were considered masculine, those that set in the evening following him — feminine, since women walk behind the man; Mercury changes sex in each case. The planets' position relative to the horizon can also affect their sex, for when they rise in the East they are full of humidity, transforming themselves into masculinity as they rise and dry out in the heat of the Sun, relapsing into womanhood as they set in the West (B-L, 102). They can change their sex according to the sign of the zodiac they inhabit, an important point for the astrologer, as Bouché-Leclercq indicates, who would otherwise have had only two female planets to account for the sex of children in a geniture. Masculine planets, finally, are diurnal, associated with the day, sunlight, and work; feminine ones are nocturnal, night — and presumably bed — being the domain of women. While these anthropomorphic categories obviously expressed the male dominance so familiar in Greek thought (see Lloyd 1961, 1983; Vickers 1973), we should note that astrology imposed its own categories onto its cultural inheritance. The need to balance out the two groups, nocturnal and diurnal, made the astrologers, "ready to sacrifice anything for symmetry,... break the logical correspondence between the masculine and the diurnal. Mercury being set on one side as a variable, four planets had still to be divided between the sun and the moon, three of them masculine." Symmetry was achieved not by transferring Saturn to the ranks of the women, which might have been justified in terms of physics (he being cold and humid), but by relegating Mars to the nocturnal sphere on the grounds of mythology, that is the attractions of Venus (B-L, 104–105). This had the added advantage of putting one maleficent planet in each group. But once again categories are collapsed or conflated arbitrarily, to fit some overriding need.

This brief account of the treatment of the planets has already revealed some of the methods of astrology. Human-based categories — temperament, sex, a disposition towards friendliness or hostility — predominate. Ascribed qualities then form the basis of a classification system in which the needs of the matrix or set that the astrologer wishes to establish can override other factors. Reasoning from analogy takes the place of observation, or logic. All these characteristics are seen in much greater detail in the elaboration of the zodiac or *Tierkreis*. This term describes the zone of the heavens on either side of the ecliptic, or apparent path of the Sun at an angle of 23.5° to the equator. The name comes from the signs (*Zōdia*) of creatures, those imaginary figures discerned in the constellations through which the Sun passes. To astronomers, Bouché-Leclercq writes, these signs are only words, but for astrologers they are concrete realities, living, powerful (B-L, 130). To many people the zodiacal signs, like the months of the year, are just given, existing as of necessity and from time immemorial. Yet historians of astronomy and astrology — such as Franz Boll, Carl Bezold, Wilhelm and Hans Georg Gundel, Robert Böker, Otto Neugebauer and B. L. Van der Waerden[11] — in reconstructing the stages by which the zodiac took shape (a history in which there are inevitable gaps, and much disagreement), enable us to see these images and the systems deriving from them as neither inherent nor necessary, but man-made, variable, and arbitrary.

To begin with, an observer detected a certain pattern or form in a group of adjacent stars, saw in it a resemblance to something from our human world, and named it accordingly. In the testimony of Geminus of Rhodes (a Stoic philosopher and pupil of Posidonius, first century B.C.), in his *Eisagoge* or *Introduction to Phenomena*:

> By zodiacal sign we understand two things: first, a twelfth part of the zodiac, that is, a space of a certain extent defined by stars or points; secondly, a picture derived from the stars according to similarity and position.[12]

The stars composing this form vary in size and strength, as one might expect, but also, disconcertingly, in number. The number of stars composing Virgo is 19 or 20, according to some early sources, but according to Ptolemy it comprised 32; Sagittarius was made up of 31 stars to Ptolemy, but of only 16 according to Hipparchus (*Z*, 474). The pictures were made up by a process of "joining the dots": thus, Taurus is composed of the Hyades, in which those ancient observers could perceive the stylized triangular shape of the bull's head, especially its left eye (the star Aldebaran; *Z*, 471). H.G. Gundel classifies Aries as a "künstliches Sternbild" (invented star-picture) (as he does for Cancer, Virgo, Libra, Sagittarius, Capricorn, and Pisces), but finds Taurus a "natürliches Sternbild" (natural star-picture) (as he does for Gemini, Leo, Scorpio, and Aquarius). Yet the truth is they are all "künstlich," made by the art of man: Cancer was also known as the Ass (Asinus) or the Manger (Praesepe), depending on how you looked at it. One zodiacal sign, Libra, was established rather late. Whereas the zodiac became established in Greece in the fifth or sixth century (disputed — Van der Waerden credits its introduction to Cleostratus, c. 520 B.C.: *Z*, 489), Libra was designated only in 237 B.C., having previously been seen as the claws of Scorpio (*Z*, 473).

> Scorpio once occupied so much more than thirty degrees that to get a duodecimal zodiac it had to be made into two signs; and the two legs of Ophicus protruded, and still do, right into the zodiac and even south of the ecliptic. (Goold, in his edition of Manilius, p. xxii)

The fact that signs could be made or seen at this late date proves that they were all human constructs.

The elaboration of the zodiac brought astrology into some conflict with astronomy. Greek astronomers held that the seven planets did not revolve in the same sphere as the fixed stars of which the zodiac is composed, but in separate spheres between them and the Earth. But astrology had to be able to interrelate its categories of stellar influence, and did so in defiance of astronomical theory and observation. As Hans Georg Gundel wrote, astrologers had

> the naive concept that a planet 'stands' in the zodiac or in a particular part of it. This idea, which was and is fundamental to astrology, maintained itself in the face of the highly differentiated cosmic systems of antiquity and despite frequent contradictions. (Z, 477, citing Hyginus, *Astron.* IV, 14)

More striking, since the zodiacal constellations belong to the fixed stars, they take part in the apparent circular movement of the stars from East to West. The Greek astronomer Hipparchus discovered as early as 160 B.C. the phenomenon known as the "precession of the equinoxes," "by which the zodiacal band changes its position relative to a given point on the ecliptic by one sign in about 2,000 years" (Shumaker 1972, 15). The result is that by now the zodiac is displaced by about one whole sign, so that Aries has moved into the place of Pisces, Leo into the place of Cancer. Origen, an early Christian critic of astrology (late second century A.D.), noted that the discrepancy between the zodiacal signs and the star pictures would gradually make the prophecies of the astrologers illusory, since they would be dealing with displaced and imaginary values (Boll, Bezold, and Gundel 1926, 131). Yet astrologers have not been dissuaded, preferring an imaginary "fixed" zodiac to a real, moving one. In other words, the zodiacal signs have persisted as a detached and self-contained system, independent of the original observational element.

Another discrepancy between systems was created by the imposition of a geometrical model onto the mythological-anthropomorphic form of the zodiac. At some point in the Hellenistic period, the purely formal geometrical division of the circle into 360° was superimposed onto the zodiac, dividing the signs into 12 equal sectors of 30° each. The striking contradiction between this neat formal division and the observably unequal nature of the constellations was pointed out as early as the first century B.C. by Geminus of Rhodes. To continue the passage quoted earlier:

> As far as the zodiacal signs are concerned (in the first sense, of pure

> space) they are all of the same size, since the zodiacal circle is divided into twelve equal parts by means of observational instruments. On the other hand, the signs in the sense of star-pictures are not of equal size, nor composed of an equal number of stars, nor do they completely fill the space allocated to them. Some are of smaller extent, such as Cancer, who takes up only a little part of the space assigned him; while others, such as Virgo, exceed it and even take up part of the signs preceding and following. Further, some of the twelve star-pictures do not even lie within the bounds of the zodiac, but stretch to the north of the band (as Leo), or to the south (as Scorpio). (Z, 481)

Hipparchus had already pointed out the discrepancy between the equality postulated by the geometrical division and the actual visible constellations (Z, 482). But astrology went ahead with the geometrical system — which Bouché-Leclercq described as "a violent wrenching of nature" (B-L, 62)[13] — because of the great benefits it offered in terms of partition and computation.

The implied symmetry of the division of the zodiacal signs into equal sectors was belied again by the rising and setting of the parts of the zodiac. Hipparchus pointed out in the second century B.C. that the irregular placing of the constellations north and south of the ecliptic produced great variations in the times of rising and setting compared to the geometrical model (Z, 483). Geminus showed that signs that rise when the ecliptic is steep "take longer to rise, but are quicker to set," and vice versa (Z, 484). He also drew attention to the discrepancy between the symmetry of the geometric model and the inequality of the Sun's apparent course through them: "All the signs are segments of 30°, but the Sun does not run through them all in 30 days" (Z, 486). It is due to the division of the circle into 12 equal parts that we owe the increase in the number of signs from 11 to 12 by the subdivision of Scorpio. At all events, the number 12 played a considerable part in further elaborations of astrology, into 12 gods (Z, 545), 12 apostles, 12 patriarchs, 12 angels in paradise (Z, 547).

II

Turning now from the planets and signs, with their man-made forms and attributes, to the systems used to interrelate these prime categories, anyone who works at all extensively on the occult sciences will soon see that one of their great obsessions was classification. Bouché-Leclercq is an eloquent commentator on this process in astrology, recording that Pythagoreanism provided astrology with "an inexhaustible mine of combinations"[14] of which the astrologers "have merely limited the number of computable combinations and disqualified certain types" (B-L, 7, 9).[15] He describes a "hybrid combination" (B-L, 83),[16] records the "minute precision"[17] with which horoscopes were assembled (B-L, 51), and alludes to the "game of Patience"[18] with which astrologers elaborated the doctrine of the decans (B-L, 215). In one sequence he surveys the whole process of the divisions or

> segmentations that the astrologers multiplied at will in order to adorn every point on the sphere with diverse influences among which they could choose. There were all sorts of combinations, consisting of the assignment of planets among the signs,... or even of signs among the planets, variously arranged, sometimes following the "natural" order of the trigons, and dissecting the signs into ever smaller parts — thirds, sevenths, ninths, twelfths, even thirtieths, i.e., ending up with a single degree of arc. (B-L, 216)[19]

Reading this classic study, one is struck by such frequently recurring words as "combinations," "correspondences," "distribution," "repartition," share," "distribute," "attribute," "associate," and "attach."[20] It shows how astrologers could not resist the temptation to use all possible permutations (B-L, 241), how one system "was still not complicated enough for astrologers seeking reasons and especially excuses" (B-L, 246),[21] a complexity of elaboration that sometimes resulted in chaos (B-L, 244, 254). As Bouché-Leclercq wryly puts it: "the imagination of astrologers is inexhaustible; those who frequent them must have a goodly store of patience" (B-L, 294).[22] Their permutations proliferate beyond human patience (B-L, 306–9), and are applied without system or logic (B-L, 298, 316–19, 405–6, 428, 478): the correlation of the signs of the zodiac with the parts of the body being "a system put together by veritable cloudmakers" (B-L, 321).[23]

Bouché-Leclercq intersperses his marvelously sustained analysis with flashes of mordant humor. One account he summarizes "is not, be it admitted, a masterpiece; but it is adequate to give us an idea of what a virtuoso could have done with the same pieces on the chessboard, by introducing into the analysis"[24] a whole new set of variables (B-L, 474). Chief among these virtuosos of classification and rearrangement were the Egyptian priests with their systems of "chronocratories" or time-gods:

> Fertilized by personal interest and by leisure, their imagination had to produce various systems of chronocratories regulating the years, months, decades, days, and hours: some regular and relatively intelligible, others irregular, unintelligible, and for just that reason viewed all the more as revealed mysteries. (B-L, 478)[25]

Worked out in detail, taking into account each planet's rule as chronocrator in relation to which succeeds and which follows him, "the number of possible arrangements becomes dizzying" (B-L, 494).[26] When zodiacal chronocrators are correlated with planetary chronocrators, a "baroque combination"[27] results and Bouché-Leclercq rejoices at being able to leave one of the astrologers "in his predicament, and finally extricate ourselves from this arithmetic hell"[28] — only to fall into another (B-L, 497).

Yet, for all their madness, this proliferation of systems had a method. Given that the aim of astrology is to correlate the qualities and attributes of the stars — category A — with the birth, life, and fortunes of humans beings — category B — then the more eventualities the system offers, the greater the possibility of finding suitable correlations between A and B. Sometimes the desire for classification seems to exist as an end in itself, with no immediate application to human life. (We should, in passing, remember that astrology was codified in the Hellenistic period, which saw a similar proliferation of categories in the handbooks on rhetoric.) Thus the signs of the zodiac were arranged into all possible categories. As Hans-Georg Gundel sums up this vastly complicated process:

> Just as with the planets the star-pictures of the zodiac were classified from various aspects into categories, in order to yield principles for astrological verdicts. Names, myths, external form and nature, extension and place in the ecliptic, relation to the equinoctial points, the way in which the signs appear over the horizon (whether head-first or feet-first), the number of stars and their brightness — all these criteria open up as many possibilities for grouping the zodiacal figures into categories as do their relationship to the sun's course, the seasons, the winds, or the weather. Study of the source material for these doctrines soon makes one conclude that such classifications are neither impartial nor follow any logical order.... (Z, 548)

So the zodiacal signs can be classified according to their anthropo- or theriomorphic shape: humans, animals and mixed; animals with horns or with many heads; land animals, water animals, and amphibious animals (Taurus); those with four or more feet; those that walk, swim, fly, creep; large or small; wild or tame; sexy or chaste. They can be classified according to their posture in the sky, standing up, running, sitting, or lying. They can be linked to the four seasons or divided into male and female. Here the role of numerology as an organizing element in many of the occult sciences becomes apparent, for in the Pythagorean tradition uneven numbers are male, even numbers female. Beginning with Aries, then,[29] and simply alternating the ascription of sex, we have as masculine

> Aries, Gemini, Leo, Libra, Sagittarius, Aquarius; and as feminine — with the connotation of passivity — Taurus, Cancer, Virgo, Scorpio, Capricorn, Pisces. (Z, 550)

Sex can also be assigned according to the quadrant the signs occupy, or according to whether they are diurnal and masculine, or nocturnal and feminine. This whole system is of great importance for birth horoscopes, as a male child born under a female sign might be thought to be at risk from hostile women.

These types of classification are still concrete, visual, anthropomorphic. From the shape or sex of the sign (category A), further significances for human life (category B) are derived. Another group of classifications works on a more abstract plane, using simple operations in geometry and arithmetic, but still deriving anthropomorphic data from these mathematical operations. Bouché-Leclercq describes one of these processes as "that mystico-mathematical science" (B-L, 82),[30] meaning, I suppose, that it confuses mathematical processes with symbols and analogies arbitrarily composed. It should be stressed, however, that the goal of astrology in this period was not at all mystical, but earth-bound and individual-centered. Whereas a scientist approaches physical or natural phenomena directly, with an interest in the things themselves — what are they? how do they function? what method must I use to understand them? — the occultist, or rather his client, approaches phenomena within an anthropomorphical and analogical frame in order to ask: will I be happy? will I be rich? healthy? successful? And if these things will not come to me automatically, how can I use the stars' influence to achieve good things, or ward off evil? Astrology refers always to individual fortune, not to science as a system developing and extending knowledge of physical reality. In terms of its *telos*, then, the whole amazingly elaborate system, with its proliferation of classes and categories, was designed to create enough reference points in the lore of the stars to accommodate all possible human situations.

In these more abstract classification systems the determining factor was no longer the shape of the constellation (*Tier oder Sternbild*) but its sign, and the place of that sign within a 360 degree circle. That figure can be divided in many ways, for instance into 360 parts of one degree each, into the *monomoiriai* or *Schicksalsmächte*, the astral beings thought to control each degree. Gundel analyzes four variant systems of dividing the signs according to their houses or trigons, including one in which the individual signs are ignored and the powers of fate are attributed to the full 360 degree circle, "that is, to wholly imaginary dimensions of space and quantity," and concludes: "Consistency is the last thing one can expect from such games of patience ..." (*Z*, 564). Another system actually divides up the shape of the zodiacal figures into parts, that is, the horns of Aries, then its head, body, and so on: "In defiance of the whole iconographic and astral conception, each constellation is forced into the 30 degrees allotted it, and then chopped up," this for use by astrological medicine (*Z*, 515). Other occultists or sects distinguish 360 worlds, 360 gods, 365 types of evil, 365 heavens, 365 seeds in the embryo. Or the degrees can be divided into their minutes, giving 60 for each degree, 1,800 for each 30 degree sector, or 21,600 parts for the whole zodiac (*Z*, 565–66). Given the popular belief that the skies revolve each day through 360°, then at each degree a deity is in the ascendant, that position of the greatest influence over the horoscope. In such a way "the number of explanatory possibilities is increased and the interpreter's imagination is given a large field to play in" (*Z*, 566).

One of the most influential forms of division applied to the circle was to divide each sign into three, giving 36 decans, each presiding over 10° of the circle. As Gundel has shown, this sytem derived from Egyptian divisions of the hours of the day and the days of the week. To each decan were attributed a name, a form and a nature, and from this ascribed personality further deductions were made according to their position in the horoscope. Once again, anthropomorphic symbolic significance is read into a classification system, and read out again in various combinations. The twelve signs could be linked to various divisions of time, such as the Egyptian division of the day into 12 double hours (the *dodekaoros*), or the related sequence of 12 days, 12 months, 12 years (*dodekaeteris*). In this *chronokrateia*, or rule of the time gods, the place of the Sun in the horoscope determines the first year of a child's life, that of the Moon the following 12 months, days and hours, and so on: "by the simultaneous interweaving of these cycles the interpretative possibilities are enlarged until explanations can be provided for every change in human life" (*Z*, 566–67). The 12 zodiacal signs could also be divided each into 12 parts of 2.5°, where each zodiacal sign predominates in turn in the smaller division, giving 144 *dodekaeteria* and providing the star-interpreter with even more variations. Once again, the mathematical operation is at the service of the overriding anthropomorphic symbolism, for the presence of a sign in each unit could be used to strengthen or weaken the dominant sign, to determine sex (by alternation), and to affect the general tendency towards those Manichaean alternatives of good and evil fortune.

> The friendship and enmity that exists between rulers and co-rulers, the whole passion-loaded game of love and hate among the astral powers, who send human beings into the world and imbue them with guilt, this antagonism continues to work itself out on earth in the lives of children born under these stars.... (*Z*, 561)

As well as using simple geometrical and arithmetical processes to isolate and empower specific zodiacal signs or their subdivisions, astrology applied these processes to the relationships between the signs, again drawing from the anthropomorphic qualities already attributed to them a further series of implications for human life. If we recall that the signs can be thought of as masculine or feminine, active or passive, friendly or hostile, then combinations of these qualities can be auspicious or inauspicious. (It is curious how astrology radically simplified the whole of human life into the alternatives of good or evil fortune: it turned a continuum ranging from "more" to "less" into a dichotomy of "yes" or "no.") This doctrine was built up by placing the zodiacal signs around a circle, sitting in a ring looking at each other, an idea that makes more sense if we recall that for the astrologers "the zodiacal signs were living beings" (B-L, 159).[31] The first system placed parallel lines across the circle (with a little bit of maneuvering, as Bouché-Leclercq showed: B-L, 160ff.), creating *antiscia* between the signs, which were then divided into those that "command," those that "obey," and those that "listen to" each other, a distribution that created much confusion and ambiguity (B-L, 163–64). The second system, that of the "aspects," made configurations between the signs by effecting all possible links across the circle with a quotient of 2 or 3. In these *Polygonalverbindungen*, as Gundel puts it, "here again the fictitious basis for the geometrization is provided by the specific emotional states and dispositions of the astral beings, as expressed through their opposed positions and emanations." (Z, 533)

The first designated links are between diametrically opposed signs, 180° apart, known as oppositions. These were generally regarded as unfavorable (B-L, 167–68; *Z*, 555–56). Secondly, signs could be linked up at a 120° angle, in groups of three, or trigons. This was deemed to be "the most effective and favorable figure" (B-L, 169; *Z*, 556–57),[32] since it joined signs of the same sex, between which there could be no hostility(!). Thirdly, zodiacal signs at 90° to each other could be linked, giving a tetragon, which was considered "antipathetic and unfavorable"[33] according to Ptolemy, because it united signs of both sexes — a source of discord — and because signs of the same sex were placed in opposition to each other (B-L, 170–71; *Z*, 557–58). Luckily, the last possible combination, into 2 hexagons, once more linked signs of the same sex and was considered to be "favorable and auspicious" (Z, 558; B-L, 171–72). The authorities on ancient astrology agree that in this area, above all, number mysticism could be indulged in.

> The theory of aspects, the most admired creation of scientific astrology and the basis of all its computations, ultimately rests on the occult properties that the Pythagoreans accorded to numbers and geometrical figures — properties dwelt on *ad nauseam* by all mystics and defenders of astrology. (B-L, 172–73)[34]

Gundel says that this doctrine became "an exercise ground for the formulation of Pythagorean number mysticism in relation to cosmic harmonies" (Z, 558).

All the divisions of the zodiac so far considered have been spatial, based on geometrical principles. Another spatial division, of immense importance in the history of astrology, was spatial but not geometrical — the doctrine of "houses." This derived from the attempt to interrelate the seven planets and the 12 zodiacal signs, imagining that each planet would prefer one sign above the others, would feel more at home in a specific place on its journey. This whole system was open to the objection formulated by the author of *Hermippus*: "how can a planet, that by definition never stops moving, make himself at home somewhere?" (B-L, 183). Obviously, the astrologers who evolved the theory of the houses were imposing one category on another in order to derive some hierarchy of preference or influence. There was a basic numerical problem of correlating two unequal numbers, 7 and 12, and perhaps for that reason the Sun and the Moon were omitted, and the remaining 5 planets were assigned a nocturnal and a diurnal position (B-L, 186–92). As Ptolemy explains in the *Tetrabiblos* (I. 17), "the planets have familiarity with the parts of the zodiac." The two most northerly zodiacal signs are said to be nearest the two luminaries, which gives the correlation

Sun: Leo — masculine
Moon: Cancer — feminine

The astrologers who preceded Ptolemy "assumed the semicircle from Leo to Capricorn to be solar and that from Aquarius to Cancer to be lunar," assigning the following signs to the planets:

Saturn: Capricorn, Aquarius
Jupiter: Sagittarius, Pisces
Mars: Scorpio, Aries
Venus: Libra, Taurus
Mercury: Gemini, Virgo

Again a numerical category was allowed to determine a supposed relationship between the stars. The astrologers, Bouché-Leclercq writes, "were most ingenious people, and were not at all embarrassed to justify a system that was manifestly imposed upon them by the tyranny of numbers" (B-L, 187).[35] Further, especially in the system of Firmicus, "the extravagance lies in confounding the real presence of a star with its house, in supposing that the planet, moving from one house to the next, is in diametric, trigonal, and quadratic 'conjunction' with the Sun or Moon, *taken to be motionless*" (B-L, 187 n.; my italics).[36] Again we see how, in astrology, whenever the exigencies of a classification system collide with observed phenomena, it is the latter that must be adapted. *Res* yield to *verba*.

Such is the complexity of astrology that from time to time we must remind ourselves that is was a mental or verbal construct, using the movement or configurations of cosmic bodies as the basis for a symbol system. By the processes of attribution and anthropomorphic analogy, human qualities are first attributed to the stars and then deduced from them. All classification systems are designed to yield the maximum of eventualities, as if dividing a wall or a cabinet into the greatest possible number of compartments. The most common method of producing these compartments is by division and distribution, cutting up the image of the zodiac in different ways according to different criteria. Or two categories are placed together, superimposed, and the second read in terms of the first. One could describe this as a process of assimilation, new material being adjusted to the original matrix and taking over its qualities. We have looked at some of the spatial divisions and correlations, often in geometric terms; equally important are what I would call conceptual correlations, where one matrix or set is juxtaposed with another unit and a one-to-one transfer of the constituents or attributes takes place. Whereas the methods so far examined apply internally to astrology, with these astrology extended its influence to meteorology, geography, ethnography, physiology, mineralogy, botany, and medicine.

Once again it must be stressed that these categories are not natural, inherent, or essential, but are man-made, and subject to variation and manipulation. Typically the signs of the zodiac, or groups of them — such as the trigons — are set out on one side of a table and other categories correlated with them, forming a scheme or grid. Thus the four seasons can be correlated with the four Aristotelian qualities, elements, the cardinal points and the winds that blow from them, colors, humors, and the four ages of man. If the seven planets form the basic matrix, there may be seven colors, and seven ages of man. In other words, the subordinate category is flexible and can be adjusted to fit the profile of the dominant one. According to the constituent parts of the dominant matrix the body can be divided into 7 parts (planets); 12 (zodiac); 36 (decans); 360 (circle); 365 (days of the year). The subdivisions are variable, then, and they are also arbitrary. It is up to the individual correlator to distinguish and link up the elements as he wishes. Thus, in correlating the zodiacal signs with the parts of the body (astrological *melothesia*), one common system (represented by Manilius) was to imagine the human body stretched over the extended zodiac so that the head covers Aries and the feet cover Pisces ("the Fish," Bouché-Leclerq remarks, "which have no feet themselves, but make up for this appalling lack of suitability by the fact that there are two of them": B-L, 319).[37] Thus the neck corresponds to Taurus, the animal so strong in that part; the shoulders and hands, twinned members, with Gemini; the chest to the shell of the crab; the sides to Leo; the lower stomach and intestines to Virgo; the buttocks that keep the body in equilibrium to Libra; the genitals to Scorpio; the thighs to Sagittarius; the knees to Capricorn (the kneeling goat); the legs to Aquarius; and the feet to Pisces (*Z*, 579). Other practitioners did not agree, replacing Scorpio in the genitals by Sagittarius, shooter of Cupid's arrow, or objecting to the location of Virgo in or near the pudenda (B-L, 319 n.; *Z*, 580–81).

With the division into decans or 36 parts, an equally arbitrary assignation of the constituents took place. According to one system, in Cancer the first decan rules the arteries, the second the lungs, the third the heart, and so on. The great number of variant schemes listed by Gundel shows all too clearly the arbitrary nature of such correspondences. Similarly, when the scheme of the zodiac was correlated with lists of metals, stones, plants, animals, colors, or whatever other category, the many variations that are recorded testify to the activity of human choice and human disagreement. Yet of course these correspondences were profoundly important, for numerology and astrology played a large part in medicine up to the eighteenth century. Galen, in many respects an empirical physician, wrote a treatise on the lucky and unlucky days within sickness that affected bloodletting and other therapy for some 2,000 years. In surgery, doctors were discouraged from operating when the Moon was in the ascendant, or when a malign planet such as Mars or Saturn predominated. As for remedies, plants and herbs had their allocated place in the zodiacal scheme,and were administered according to the same rules for determining auspicious and inauspicious effects. Alternatively, the type of visual analogical thinking that we have met so often in astrology was applied to the shape, the color, and the texture of plants, giving the so-called doctrine of "signatures."

In all these correspondences, or category-fits, one matrix is deemed to determine the nature of the other sets correlated with it. That is, two or more sets displayed side by side have each a vertical and a horizontal axis, the grand assumption of the occult sciences being that such schemes can be "read" laterally, or horizontally. Qualities assigned to the original matrix are transferred or translated to those corresponding to it. Yet, as I have shown elsewhere (Vickers 1988), in practice many difficulties arise with this horizontal transfer, for there is no innate reason for an identity, or even a similarity, between the constituents. One striking instance of the variable, contingent nature of these schemes is astrological geography or ethnography. Again, the many variations in the lists correlating the signs of the zodiac or the planets with the parts of the known earth (excluding America!) underline the element of choice and manipulation in astrology (*Z*, 573–76; B-L, 327–47: a superb exposé). In Book II of his *Tetrabiblos*, Ptolemy attempts a classification of his own, more complicated than the others, and involving more factors. He begins by dividing the world into four regions or quadrants, each of which is dominated by a zodiacal trigon. The northwest quadrant, Europe, consists of Aries, Leo, and Sagittarius, and is dominated by Jupiter (north wind) and Mars (southwest wind); the southeast, Ethiopia and Greater Asia, consists of Taurus, Virgo, and Capricorn, and is governed by Venus (south wind) and Saturn (east wind); Gemini, Libra, and Aquarius make up the northeast (Scythia), which is governed by Saturn (east wind) and Jupiter (north wind); leaving Cancer, Scorpio, and Pisces to constitute the southwest (Libya), which is ruled by Mars (west wind) and Venus (south wind). Within these zones Ptolemy then combines planetary and zodiacal influences, together with the rising and setting of the sun, in a free-flowing series of deductions from the given hierarchical categories. Perhaps nowhere else in the history of astrology is its tendency towards the elaboration of self-contained systems more apparent (II. 2–3).

In terms of climate, those nations near the equator "have black skins and thick woolly hair, are contracted in form and shrunken in stature," and are mostly savage because "continually oppressed by heat." The Scythians, formed by cold and moisture, are "white in complexion, straight-haired, tall, and well-nourished." Exact antitypes in other ways to the Ethiopians, they are also "savage in their habits because their dwelling-places are continually cold." Those nations sharing neither extreme of cold or heat are "therefore medium in coloring, of moderate stature, in nature equable" and civilized. The southernmost of them, the Egyptians and Chaldeans, who discovered astrology, are inventive and sagacious. But they can be sub-L divided into the eastern group, which "are more masculine, vigorous of soul," and open, because they share these qualities with the Sun and the oriental quadrant in general, and are therefore "diurnal, masculine, and right-handed." Those to the west "are more feminine, softer of soul, and secretive" — like women — also "left-handed." The tidiness of these binary oppositions, so basic to astrology, recurs in the subsequent division of the quadrants into opposite triangles, interrelated according to whether they are at the center of the known world or its extremity. Then Ptolemy surveys the individual nations, ascribing to each the qualities of the planets. In the northwest triangle, where the trigon is Aries, Leo, and Sagittarius, the people are "independent, liberty-loving, fond of arms, industrious," and so forth. But because of the occidental aspects of Jupiter and Mars they seem to lack virility, for "they are without passion for women and look down upon the pleasures of love," preferring the company of men. As Bouché-Leclercq showed (B-L, 340, n. 2), the charge of homosexuality against the northern barbarians goes back to Aristotle and was frequently repeated: Ptolemy is using astrology to rationalize an old racial prejudice.

It is not surprising to discover that those parts situated "about the center of the inhabited world" — Thrace, Macedonia, Greece and its Islands, Cyprus — are "noble and independent, because of Mars; ... liberty-loving and self-governing, democratic and framers of law, through Jupiter; lovers of music and learning, clean livers" and "addicted to the performance of mysteries," thanks to Venus; "social, hospitable, justice-loving, fond of letters, and very effective in eloquence, through Mercury." The transparent prejudice in favor of one's own people and distrust of foreigners is justified at each point by astrological considerations. In the southeast triangle Venus and Saturn govern, so those people are "ardent, concupiscent, and inclined to the pleasures of love"; they like dancing, and live luxuriously. Whereas the northern barbarians were homosexual, these southern Orientals — again following an ancient Greek tradition accusing them of sexual immorality and incest (B-L, 341, n. 2) — have sex with women "openly and not in secret," and "most of them beget children by their own mothers." Some nations are "luxurious and clean," but those in India "have familiarity with Capricorn and Saturn," and are thus "ugly, unclean, and bestial." Those around the Dead Sea are gifted in trade, but "are more unscrupulous, despicable cowards, treacherous, servile, and in general fickle," largely due to the presence of Mercury. The Jews are "bold, godless, and scheming," another preexisting Greek prejudice against Jewish monotheism and disregard of pagan gods.

Much more fortunate are those in the northeast, governed by Saturn and Jupiter in oriental aspect, who are cleanly, learned, just, liberal, lofty, noble, and magnanimous. However, those nations familiar with Mercury are "inclined to rascality," those in familiarity with Aquarius and Saturn are "more ungentle, stern, and bestial." Given the two impulses at work here, Greek xenophobia and planetary determinism, most of Ptolemy's deductions are predictable. Where the Moon is in the Orient, and therefore masculine, the women are "virile, commanding, and warlike, like the Amazons" (B-L, 149). East and West forming a binary opposition means that even sexual practices will be diametrically opposed. Those governed by Venus and Saturn in oriental aspects have "the practice of consecrating the genital organs because of the aspect of the aforesaid stars, which is by nature generative" (B-L, 141). The men, however, influenced by "the aspect of the maleficent planets" — Saturn and Mercury — here "in combination with Venus occidental" are unsound and effeminate of soul, and some even hold in contempt the organs of generation" (B-L, 155). For Ptolemy, planets rising increase sexuality, planets setting diminish it. So, in discussing "Bodily Form and Temperament" (III. 11), he works out a neat series of oppositions:

Saturn orient: Men are "dark-skinned, robust, ... curly-haired, hairy-chested."

Saturn setting: Men are "dark, slender, ... straight-haired, with little hair on the body."

Jupiter rising: Men have "moderately curling hair."

Jupiter setting: Men have "lank hair or even bald in front and on the crown."

Mars rising: Men have "thick hair, somewhat curly."

Mars setting: Men have "not much hair on the body, and straight yellow hair." (B-L, 309–11)

In this sequence Ptolemy correlates two categories: planetary and zodiacal movements (A) with human mores, sexuality, religion, and physiology (B). Who can say that the relationship between the two categories is anything other than analogical and circular? To Mercury is ascribed deviousness, instability: those familiar with him will be crooks and cheats. When planets rise, their influence is strong and favorable; when they set, their powers wane, and so do those of the humans unfortunate enough to be associated with them. In a way the whole system is tautological, for it merely reaffirms the binary preferences basic to Greek culture: as right is better than left, as male is better than female, as east is better than west, as one's own countrymen are better than foreigners, so all the planets affected by these dichotomies will work on us, accordingly, for better or for worse. The categories of astrology, then, are neither natural nor physical but cultural, anthropological, anthropomorphic. We are dealing with a sign system and classification system obeying its own increasingly elaborate rules, having totally lost contact with the physical reality that it still pretends to depend on and interpret.

III

Whatever our judgment of astrology as a science, there is no denying its success as a quasi-religious belief system. If, as Clifford Geertz has said, one of the main tasks of a religion is to make sense of the world we live in (Geertz 1956), the flourishing of astrology was undoubtedly due to its claims to be able to explain the accidents of life in terms of the "lot" we are given at birth, and also to know human destiny, to predict the future, and if not control it, then manipulate or adapt it to our better fortune. Astrology was taken up at Rome and codified afresh, most notably by Manilius (see B-L, 543–627; Long 1982). It spread to Islamic countries, and the Middle Ages produced a number of learned astrologers, whose works were translated from Arabic into Latin at the entry points for Islamic culture, notably Spain. Such writers as Albumasar, Al-Kindi, Mashallah, Alchabitius, Abraham Ibn Ezra, and Albohazen Haly filius Abenragel were known all over Europe, and their books were circulated in Latin or in the vernaculars, up to the seventeenth century (see Thorndike 1923–58). The number of medieval and later manuscripts containing astrological works is enormous, and, proportionally, the number of printed books hardly less. One bibliography of books available to an educated Elizabethan runs to over 400 titles (Parr 1953). Popes and rulers took the advice of astrologers before setting their day for inauguration or coronation, or laying the foundation stone for a city. People from all stages of society had their horoscopes taken, and followed some of the teachings of astrology. Ideas and images of the stars are found throughout literature and art, right through the whole Renaissance period.

Yet, by one of those striking paradoxes in the history of thought, astrology was always vigorously criticized. In antiquity there were Carneades, Cicero, Sextus, and Empiricus (Long 1982); among Church opponents were Justin, Origen, Gregory, Jerome, Ambrose, Chrysostom and St. Augustine (Amand 1945); in the Middle Ages critical evaluations were made by Al-Farabi, Ibn Sina, Ibn Haldun, Nicholas Oresme, Henry of Hesse, and others (Madkour 1969; North 1980; Coopland 1952; Pruckner 1933; Wedel 1920; Lauer 1966). None of these attacks, however well-argued or polemical, succeeded in stopping astrology. Indeed, as Richard Lemay had said, the only effect of the attacks was to produce more works written in its defence (Lemay 1962, xxxvii, n.). The same holds for the Renaissance: the attacks by Pico della Mirandola and Savonarola in Florence at the end of the fifteenth century yielded several defences by Bellanti, Pontano, and others (Shumaker 1972, 27–41). The polemical books in Elizabethan England by William Fulke, Henry Howard, and John Chamber produced a whole series of defences, and the seventeenth century saw a long-drawn-out running battle between opponents and defenders (Allen 1941; Kocher 1953, esp. 201–24; Dick 1944; Thomas 1973, ch. 10–12). Yet at this point astrology began to cave in. It is generally admitted that by the end of the seventeenth century astrology was discredited as a science, and came to be associated with quacks, rogues, and charlatans. Astrological remedies even disappeared from medicine.[38]

Various explanations have been put forward for this eclipse of astrology — the rise of rationalism, the emergence of the new sciences, the corruptions of the practitioners themselves, the role of astrology in creating disorder in religion and society. One source that has yet to be analyzed in any depth, curiously enough, is the actual critique of astrology produced between 1490 and 1690, and although I would not suggest any direct causal link between these criticisms and the subject's eventual decline, they are important documents that deserve closer study.

The growth of the occult sciences in the Renaissance is a phenomenon that has attracted much attention. In a sense one could say that Florence in the late fifteenth century resembled Hellenistic Greece and Egypt, the birthplace of the occult in the Western world, in being a meeting-point for an enormous range of philosophical and religious traditions. A learned scholar such as Ficino or Pico della Mirandola would have been exposed to Christian theology, Aristotelian logic and physics, medieval scholastic interpretations and syntheses of Aristotelianism and Platonism; Plato's own works; Neoplatonism, especially Plotinus and Porphyry; Pythagorean numerology; the Cabala; the Hermetic books; the Gnostics; astrology, alchemy, and natural magic. All these traditions can be recognized in Ficino, and were passed on by him to writers who not only absorbed his work, but went back to his sources, added others, and made a new — and often larger, or more detailed — synthesis. The occult tradition has always been eclectic, and in such Renaissance occultists as Henry Cornelius Agrippa, Paracelsus, Giambattista della Porta, Oswald Croll, John Dee, Robert Fludd, and Athanasius Kircher, many ancient traditions fuse. Renaissance occultists elaborated and enlarged on what they found, so that the whole occult system, with its secrecy and obscurity, its analogical thinking, its anthropomorphic symbolism, its "correspondences" and similitudes, flourished anew, and was extended into new fields.

Yet, such was the persistence of its mental assumptions and thought processes that we find the same old methods being applied to these new areas. The correlation of extant categories (what I have called "conceptual category-fit"), often organized on a hierarchical system, reached, in the work of such learned and diligent writers as Agrippa, Fludd, and Kircher, a degree of systematization that still astonishes us today. All of life, the whole cosmos, it seemed, could be contained in one large picture or series of bracketed tables. The correlation of spatial categories was very popular. The seven planets could be superimposed on the human face, to yield a new discipline known as metoposcopy (see Cardanus 1658; King 1973; Müller-Jahnke 1982). Once again the correlation of the planets with another zone or area meant that the anthropomorphic qualities already ascribed to the planets were just read into the new space, or out from it. The seven planets could be superimposed on the human hand, yielding the new science of chiromancy. Analogical thinking vastly elaborated the doctrines of "signatures" (see Porta 1588; Croll 1609; Arber 1938; De Virville 1957; Leclercq 1918; Schneider 1950; Quecke 1955; Jüttner 1971; Müller-Jahnke 1984), and plants were categorized for medical use on the basis of their color — yellow leaves or flowers are good against jaundice; texture — prickly leaves or thorns, if eaten, are good for stomach upsets; or shape — plants that resemble teeth are good for dental treatment. Human, corporeal analogies were even applied to the testing of urine in the work of Paracelsus and Leonhard Thurneisser (see Pagel 1958, 192ff.; Bleker 1976). However, it would be a grave anachronism to describe them in any way as forerunners of modern chemical analysis. Their mental world was still one of symbols and correspondences. Thurneisser, following Paracelsus, recommended making a retort in the shape of a human body, filling it with the patient's urine, heating it, and then watching where deposits occurred as it evaporated. If they settled around the middle, then the complaint must be located in the patient's stomach, and treated accordingly. Robert Fludd noticed that urine varies in color, but he went on to interpret these visible qualities in terms of medieval color symbolism (Godwin 1979, 64–65).[39] It was several centuries before urine analysis was used properly in medicine, and by then people were no longer thinking in terms of analogies and symbols.

IV

The criticisms made of astrology during the Renaissance constitute the first coherent and systematic attack on the mental world lying behind the occult sciences. Two of the most common grounds for disapproving of astrology had been made by the Christian church for centuries, and indeed by Roman critics before them. These criticisms were that seeking to know the future through the stars was the sin of *curiositas*; it interfered with God's activities, and challenged his providence. Secondly, that a belief in the future being predestined by the stars would lead to determinism, deny God's grace, and plunge men into fatalism and apathy. I shall not discuss either of these topics, although they occur in every Renaissance attack on astrology (Amand 1945; Allen 1941). Nor will I discuss the charge that astrologers had brought upon themselves, namely, that they created social and political unrest (see Allen 1941, Thomas 1973, Larkey 1935). This development came from medieval Arabic sources, with their belief that the great conjunctions of the planets signified chaos and disorder (North 1980). The prognostications predicting the end of the world, or a universal flood in 1524 or 1586, did indeed arouse widespread panic, followed — when they did not materialize — by reactions of bitterness, frustration, and ridicule.[40]

But to leave aside the social or religious effects of astrology, Renaissance critics made a new and cogent critique of astrology as a system. They argued that it did not work: that its predictions, whenever they could be tested, turned out to be wrong; that the empirical observations which would support its claim to be a science were not obtainable, or were not accurate enough for the elaborate computations or the crucial deductions which depended on them; that it was confused, full of internal contradictions, unsystematic; and that it used man-made categories which were arbitrary and a priori, imposed on experience, not derived from it.

The biggest and most important of these attacks was by the talented disciple of Ficino, Giovanni Pico della Mirandola, in a work written in about 1493–1494, and published after his death by his nephew, Gianfrancesco. The *Disputationes adversus astrologiam divinatricem* (Bologna, 1496), fill, appropriately enough, 12 books.[41] Pico was a humanist, that is, a student of the human sciences — philosophy, logic, rhetoric, grammar — and he lived up to the standards of accuracy in textual criticism and quotation that had been established in the tradition from Petrarch to Lorenzo Valla and Politiano. He refuted astrologers who accepted a false ascription of a work to Hippocrates or other authorities (Pico, I:64f.), or were misled by a mistake in a manuscript copy, or a verbal misunderstanding (Pico, I:92, 96, 100f., 104, 274, 572; II:280, 302). This critical spirit, with its emphasis on truth and accuracy, led him to read the work of some thirty astrologers: Greek, Latin, medieval Arabic, medieval European, and some contemporary Italians (I:15ff). Some of the astrological manuscripts he worked from are still extant in Florence, and whoever looks at them even superficially will get an idea of the difficulty of the material with which he worked. The *Disputationes* is a masterpiece of humanist analysis, penetrating to the methodology of astrology, juxtaposing the authorities on many fundamental topics, and revealing a host of contradiction and confusion. Among the recurring themes of Pico's work are the traditional attacks on astrology for the vice of *curiositas*, man meddling with God's design in a way that suggests the devil's influence;[42] for encouraging determinism and denying human dignity (although the argument from free will does not occur as often as some scholars lead us to expect[43]); for damaging virtue, prudence, activity, thus not being "useful to life";[44] and for encouraging deceit, fraud, the imposture of "charlatans."[45]

As a humanist, Pico writes from the standpoint of Aristotelian physics and cosmology. He accepts that the Sun and Moon have an influence on the Earth, but only in terms of natural causes, not occult ones.[46] Pico endorses the traditional views that the heavens move in the form of the perfect circle, are eternal and incorruptible, and affect us by their "celestial heat," light, and motion (e.g., I:194–98; Rossi 1965, II:323). The progress of the seasons is caused by the changing distance of the Sun from the Earth, not by the position of the zodiacal signs (II:106). Invoking the "rudiments" of optics and philosophy, Pico reminds us that the strength or other property of a ray depends on the nature of the body emitting the radiation, and on the angle of emission, not on its zodiacal position (II:26, 56). The influence of the two main planets is universal and general, and universal causes cannot produce individual variations.[47] Everything below the Moon is subject to change, but since the planets are above her they cannot be mutable, as astrology claims, changing their nature chameleon-like according to their position vis-à-vis the zodiacal signs or other planets.[48] Where change occurs it is inherent in the property of matter itself, not caused by the position of the zodiac.[49] Pico denies that the other planets have any noticeable effect on man (I:242). As for events on earth, they are the product of human causes, not celestial: man's greed and ambition (I:378, 380, 478, 484, 492). Miracles come from miraculous causes, natural events from natural causes (I:506). A recurring response is the flat denial: *negamus* (I:368), *non possumus* (I:546), *nescio* (I:570).

Pico's most effective attacks on astrology are of two kinds. In the first he claims that astrology is not true in what it says about physical reality. Here the two Aristotelian categories he invokes are (a) truth to experience — that is, our observation of the physical world with our own senses; and (b) agreement with reason — that is, logical consistency in argument and method. Astrology is found wanting on both counts. It is inaccurate and careless: the almanacs are unreliable, and astrologers who do not know their latitude cannot determine the planets' radiation (I:156, 158, 160). Its predictions seldom come true, as everyday experience shows. Pico observed the weather predictions over a period of 130 days: they were right only six or seven times. Friends and relatives of his had predictions that turned out to be completely false. His brother-in-law died in the year that a long life had been predicted for him. His niece's husband died suddenly, having also had a long and happy life foretold.[50] The predictions of sailors, doctors, and farmers are much more reliable, since they are based on the direct experience of natural causes, whereas the astrologers study signs that are not signs, causes that are not causes (I:284, 356f.). Pico accepts the testimony of the senses as adequate guides (I:294; uncritically, perhaps, by later standards), but he invokes a perfectly coherent concept of experience. Astrologers claim that their system is based on experience, but Pico defines experience as being "founded on multiple and frequent observations of the same phenomenon." When a phenomenon recurs many times *(saepius iterata observatio faciet experimentum)* repeated observation will give the experiment (the Renaissance used the terms "experience" and "experiment" interchangeably[51]). Yet the astrologers say that the exact situation of the stars never repeats itself, or if it does, only after thousands of years. And since it cannot have experience of future life, astrology cannot claim to be based on experiment (I:476).

Astrology has an observational element but, Pico writes, it is inherently prone to error. Astrologers cannot establish time accurately, and therefore cannot be precise with their horoscopes (II:10, 154). The position of the stars is supposed to be crucial at the genesis of life, but at what point does life begin? Is it the union of the man's semen with the woman's secretions? Ptolemy recommended that the astrologer discover the time of emission of the semen, but this cannot always be ascertained, and therefore the astrologer chose the hour of birth as being more practicable (II:156, 160). But even so he could only make the right use of this if he were waiting outside the room with his astrolabe (II:288). Sundials are notoriously imprecise (II:290), and even small errors in computation have large consequences for the kind of deductions that astrology makes (II:304). When the pretended observational base of astrology is wrong, its system of signs is thrown into utter chaos (II:332); but given the innate defects of current technology it is always bound to err (II:320–24). This is one more reason for refuting astrology's claims to be based on experience.[52] Its dogmas are based not on observation or experiment but on conjecture and uncertainty, arbitrary individual pronouncements (II:126, 146).

One of Pico's most frequent strategies for invoking the criterion of reliable experience is to show up the many variations and contradictions in astrological doctrine.[53] His discussion of the lore of "houses" shows that it has no basis in reality and that the astrologers ascribe totally different properties to each "place." After a devastating analysis of their disagreements, Pico demands:

> Given such a divergence of opinion, what faith can we have in experiment when no-one yet agrees on the location of the starting or ending-points of the houses,

and each divides them up according to a different principle? It is ridiculous and false to suppose that such "places" in the sky have any inherent properties (II:34). Surveying the conflicting systems for dividing up the zodiac into decans or smaller units, Pico again asks how they can be faithful to experience (*quae fidelis et concors esse potest experientia*), for with such great variations in setting a "theme," all criteria for prediction will vary, and so the whole authority of experiment is weakened (*sic omnis experimenti auctoritas infirmatur*; II:46). The criterion of truth to experience recurs (II:72), and is used with particular force after Pico has shown that the astrologers give no less than six conflicting opinions over the sex of the signs: *quae fides erit experimenti* indeed (II:132)? Pico shows that the contradictions on major issues destroy astrology as a serious discipline (II:114). Reviewing the conflicting theories about the great conjunctions, Pico asks how one can arbitrate between astrologers who cannot show the truth by either reason or experience (II:178). The diversity of opinions over "directions," attributing the rule over certain years variously to the Moon, Mars or Jupiter, takes away all faith in experience (II:198), and reveals the vanity of a doctrine arbitrarily founded on inconsistent conjectures (II:204).

If deficient in terms of experience, astrology is just as inadequate when measured by the other Aristotelian criterion of reason. Pico gives dozens of instances where astrologers' arguments do not meet the claims of reason.[54] They are guilty of elementary logical fallacies (II:254, 376, 404, 500); they reason in a circle (II:118, 302, 450: the chapter heading of X, xv is *tum astrologos saepe uti demonstratione cirulari*); or beg the question they were supposed to prove (I:570; II:452, 506). Pico applies the dialectical technique of the dilemma, offering the astrologers two possible arguments, only to refute both of them (I:168; II:74, 236). The result of this part of his polemic is to pronounce astrology defective by both criteria, those of experience and reason.[55]

> If it could be supported by reason astrology could be declared a science; if by experiment, an art; if neither by reason nor by experiment it is neither art nor science but a temerity, an unverifiable deceit, an illusory fiction of impostors. (II:358)

It is with a certain satisfaction, the sense of a job well done, that Pico pronounces at the opening of the later books that he has cast astrology down, or destroyed its foundations before attacking its pinnacles (II:9, 357). Logic has proved a useful tool for this demolition.

In his attack on astrology's failure to meet rational or experimental criteria, Pico shows a critical spirit and sustained argumentative power that any modern scholar might envy. "What criterion do astrologers use?" he repeatedly asks (II:406, 414, 416); Albumasar is rebuked for accepting everything he read uncritically, neither refuting nor questioning (II:128), faults of which Pico was certainly never guilty. His other main strategy of refutation is to consider astrology internally, as a method on its own terms. Astrology is described as a sign system, but Pico denies that its signs have any effect on physical reality, since they are not natural but man-made.[56] He cites an astrologer who noted that the Arabs call one constellation the vulture, while Ptolemy calls it lyre-bird; this shows that the constellations are put together according to the judgment of the eye, and so the pictures made of them will differ (II:266f.). Pico dismisses the zodiacal signs as a mishmash of fantastic and heterogeneous images, based on similitudes, not the work of nature but the product of fantasy, delirium, or dreams (II:266–72). All the signs on which astrology is based are neither necessary nor innate but arbitrary (II:82, 84, 128, 134, 222). Lacking any correspondence with the real, they can have no natural foundation, and are merely mental images or concepts (II:246). These images are "vain, fantastical, the work not of God but of man, not derived from the laws of nature" but put together by a poet's imagination (II:424). These images, as he never tires of reminding us, are "fictions, fables," imaginary constructs.[57]

Pico follows up the implication of the signs being man-made, showing all the consequences of this anthropomorphic symbolism. He challenges the division of signs into diurnal/male, nocturnal/female (II:104) and objects to the mechanical way in which this division was carried out by merely alternating odd and even. With such reasoning you cannot dispute, only laugh! The proper criterion should have been the nature of the signs themselves (II:114). He ridicules the human symbolism of "left" and "right" as applied to the planets (II:108f.), likewise the use of human hierarchical categories that would make stars "strong" or "weak" (II:50, 76), "command" or "obey" (II:96) — all absurd ideas. Especially absurd is the classification of stars as malignant or beneficent according to their position (II:48). Why should Jupiter change his nature so drastically according to the company he keeps? The whole system is

> based on a certain similitude with human life, contrary to nature and to the dignity of the celestial bodies, deduced not by philosophic reason

> but with poetic license, or from pure fables and arbitrary fictions. (II:78)

The idea of the zodiacal signs or planets having dominance in turn, "as it were passing on the succession of power from hand to hand for a certain time," Pico writes, is a doctrine no doubt founded on an analogy with human customs (II:220).

As well as exposing the anthropomorphic symbolism, Pico shows how astrology depends on numerical methods, equally man-made and arbitrary. The astrological classification into just 12 signs has no basis in nature, but is merely an easy number for computation, as astrologers themselves admit (II:36, 264). The division of each sign into 30° is also not natural but human (II:38), another mark of convenience. But, he writes, the reciprocal influences of natural bodies depend on natural causes, not on numbers and figures, superimposed categories (II:52f.). As for the aspects, why is the circle only divided by 3 or 4, and not by 5 or 8 (II:p. 58)? Obviously, we answer, because otherwise the signs could not be linked up symmetrically! Similarly with the *dodecatamoirai*: planetary influence cannot work according to such man-made number systems (II: 150).

Within its various categories, Pico shows, astrology uses an enormous amount of division and subdivision. He dutifully lists all the various and often conflicting categories so produced (see, e.g., II:44, 200, 506, 510, 512, 516). *Quas omnes anatomias et dissectiones* ("what a lot of cut-up bodies") as he says mockingly at one point (II:128). He also brings out the man-made categories produced by division, distribution, and attribution, how the astrologers "refer" or "attribute" various effects to the stars (I:348, 352, and passim). But, he says, it is all conjecture, an arbitrary assignation of qualities (II:42, 128) — these are *verba*, not *res*. Above all he attacks the technique of correlating various categories with each other, showing that the supposedly horizontal interequation simply breaks down. Such analogies are "pretty" or graceful, but neither natural nor true (I:296; II:486). He rejects outright many of these correspondences.[58] Some of these category-fits fail because they attempt to correlate things different by nature (II:64) — a criticism Kepler was to echo.

So he challenges the correlation between the zodiacal signs and the four elements: "On what basis do you distribute the properties of the elements to the signs?" The traditional correlation meets no good criteria, so Pico proposes a new one, based on nature and reason (II:112f.). Pico often shows he could do the astrologer's work better than they (e.g., II:398). The consequences of these category-fits can be absurd. Saturn is the first of the planets, Aries the first sign, but they do not correspond in any way (II:370). One classification would equate Mars with Saturn, but on the astrologers' own terms the planets are basically dissimilar (II:396); if Saturn were put into a sign where he really fits, others would have to be excluded (II:404). The result of this categorization is that dry things are said to be similar to humid things (II:406).

Pico reserves his greatest scorn for the supposed correspondence between the zodiacal signs and the parts of the body (I:372, 376). How many parts does the body have, in any case? Once again the disagreement on such a fundamental issue shows how arbitrary and unreliable astrology is. Aristotle says seven, Nicomachus ten, but the astrologers twelve. Yet even so they cannot make their correlation in any rational manner, for they place the shoulders, arms, and hands under one sign, but divide up the legs into four separate signs. It would have been better, Pico writes, to have considered the qualities of the parts, whether hot or cold, then to have linked those up, and not to have proceeded in this absurd numerical order, arousing laughter by putting the feet under the dominion of Pisces, which is the only zodiacal sign not to have any feet (II:418)!

One fundamental operation of astrology not yet touched on is the way it is built up on analogy — Pico uses the correct technical term: *parabolica similitudo*. Arguments from similarity, however, are no basis for science (II:376, 450, 454). Further, they can have damaging consequences for life, as in the doctrine of the "houses," where the individual places are distributed on the basis of similitudes. The father loves his children, the fifth place is connected to the horoscope by a friendly or triangular link, so the fifth place is given to the sons. The first place is birth, the second riches, the astrologers teach, for riches are supposed to be what everyone desires. But in fact, Pico replies, many other things are more important: the gifts of the soul and body, wisdom and health (II:378). To reason by analogy is to create fallacies (II:506).

Pico refutes point by point every doctrine in astrology. The "houses" are dismissed as absurd: how can an empty space in the sky have any physical power over us? (II:22, 34, 362). The aspects are equally silly: how can one attribute power, or sex, to triangles or hexagons — which do not exist (II:60)? This is to elevate imaginary diagrams to the status of causes. The whole system, he concludes, is like a spider's web: if you look at it from a distance, it deceives you with the size of its threads, and seduces you with the grace and novelty of the weaving. But if you examine it from close up you begin to depise it, and if you touch it, it breaks into pieces (II: 518).

V

If the world were wholly rational, Pico's refutation of astrology would have killed it outright. We may recall the complaint of Lemuel Gulliver, in the letter to the relative who had urged him to publish his *Travels*:

> Pray bring to your Mind how often I desired you to consider, when you insisted on the Motive of *publick Good*; that the *Yahoos* were a Species of Animals utterly incapable of Amendment by Precepts or Examples: And so it hath proved; for instead of seeing a full stop put to all Abuses and Corruptions, at least in this little Island, as I had reason to expect: Behold, after above six Months Warning, I cannot learn that my Book hath produced one single Effect according to mine Intentions.... (Swift [1941], 6–7)

Granted that Swift has pushed Gulliver to the brink of derangement on the fourth voyage with his identification of rationality with the form of horses (Houyhnhnms), bestiality with the form of apes-men (Yahoos), the point is still valid, and cuts in two directions. Satire as a form is unable to effect change in the world; man as a creature is unable to reform his vices — otherwise there would be no evil. Similarly, the fact that Pico's attack, far from destroying astrology, produced further defences of it, should not surprise us. Scholars have brought out some of the implications of Pico's polemic in a way that makes us realize how much else in the Renaissance consciousness would have had to change before his case were universally accepted. Wayne Shumaker writes that Pico wanted to abandon a cosmos "conceived animistically, in which planets 'rejoiced' and were 'dejected', 'looked at' each other" or mankind with friendly or unfriendly feelings, and

> substitute one in which the heavenly bodies performed quite dispassionately and without consciousness roles assigned them at the beginning by a Creator-God who allowed the evil initiated by men to cause suffering but did not place in the skies forces which would dispose them to act well or badly. (Shumaker 1972, 24)

But for such a concept men had to wait a century or more, for the universe of Galileo, Kepler, and Newton. Similarly, John North concludes his study of the controversy over the "great conjunctions" with the reminder that although specific doctrines were rejected in the fifteenth century, "there was no unqualified opposition to astrology as a whole ... for the simple reason that no really effective theory of natural causation could account for all that men wished to explain." It follows that

> Works like Pico's could not alone bring an end to astrology — for that one must look for pressures from without, for new sciences which would render astrology superfluous. But Pico's book must have weakened the resolve of many intelligent individuals who might otherwise have followed the old astrology blindly, and without any thought of subjecting it to criticism by its own criteria. (North 1980, 202, 203–4)

While we have records of several Renaissance students of astrology who became disillusioned with the subject, to my knowledge no one has found specific reference to Pico among the recantations. Yet, if one reads the attacks on astrology published in England in the period from 1560 to 1680 — works by William Fulke, Henry Howard, William Perkins, John Chamber, John Geree, William Rowland, Pierre Gassendi, John Allen and Henry More — the influence of Pico is everywhere apparent.[59] The only exception seems to be Calvin, who was either ignorant of Pico's work or deliberately couched his attack in simple terms in order to reach an unlearned public (Calvin 1560). Otherwise Pico's arguments, illustrations, his whole style of polemic, recur again and again. This is not the place to document his influence, but I should like to note two ways in which his mode of attack was reiterated and extended. One element that obviously impressed his contemporaries was the appeal to reason, Aristotelian logic providing a foundation for scientific method. In the hands of William Fulke, a leading theologian, writing in 1560, this emerges as a fully-fledged statement of the principle of verifiability:

> ... this is common to all sciences, that they may bee demonstrated. For although the principle and grounds in every arte be of such nature, that they cannot bee showed and confyrmed by things more general, and therefore it is said, that they can not be proved, yet by demonstration or induction they maye be so playnly sette before our eies, that no man neede to doubte, but that they are most true and certain. (Fulke 1560, Sig. $B_{ii}r$-$B_{ii}v$)

Once the foundations of a science are firmly laid, one can build on them. In geometry, for instance, two points are joined by a straight line; the sum of the angles in a triangle equals two rectangles.

> But by no waie is it possible, that the principles of this arte of Astrologie, may be either demonstrated or proved. Ther is no mean wherby mans witte may atteyne to so great knowledge, ther is no methode, no induction, that can maintayn truth of these propositions, whyche they take for their principles.... For by what reason are ye able to

> demonstrate or shewe, that Saturne is so hurtfull, malicious, and pestilent? By Induction? (ibid., Sig. B_{iii}v-B_{iiii}r)

Modern commentators have praised Fulke's rationalistic attitude to science,[60] and although I share their respect for him, much of the credit is due to Pico, whose influence can be seen throughout the treatise.

Pierre Gassendi, an enemy of Robert Fludd and a major source for the revival of atomism in the seventeenth century, was also indebted to Pico, and stated as much on the final page of his book, where he refers his readers to "that Divine Witt, *Picus*" and his "incomparable writings" (Gassendi 1659, 162). Gassendi condenses Pico's polemic with the same gusto in demolition: chapter X is entitled "*Their Attributes of the* Parts *destroyed*"; chapter XIV "*The* Genethliacal *part of Astrology examined, and exploded*"; chapter XI "*The Celestial* Houses *demolished.*" Gassendi resumes all the crucial stages of Pico's argument: astrology has no basis in experience; its categories are arbitrary, man-made, based not on observation but on "chymerical," "meerly imaginary" fictions. The signs, aspects, houses, have no physical influence on man, the correlation of categories is whimsical and unscientific. Where Gassendi extends Pico is by updating him, scientifically. That is, he first silently omits Pico's Aristotelian heritage, and then adds new arguments from astronomy. Pico had not mentioned the procession of the equinoxes, but Gassendi recalls that the "Asterismes" have now displaced themselves by a whole sign, bringing the astrologers' deductions into "just derision" (46–47). Where Pico had referred to the inaccuracy of the almanacs, Gassendi notes that not even the tables of Tycho Brahe were as accurate as the astrologers' system claims to be (63).

Pico is updated from recent history, too: the prediction of a universal deluge in 1524, Gassendi records, was followed by unusually fine weather, and after a prediction of a "Universal Dissolution of Nature" in 1568, "not so much as a storm ensued" (65–66). New scientific discoveries are brought in to challenge astrology: "the *Telescope* hath discovered innumerable Stars in the Firmament," which ought to mean that the whole of astrology should be reformulated (67–68). Pico's demonstration that astrology cannot claim to be an experimental science, since the identical celestial configuration recurs only after "many millions of years," is endorsed and updated:

> Further, the Ancients were ignorant of all those notable discoveries, that have been lately made as well of the Fixt, as Erratique Stars; which would have required them to alter their Astrological decrees; and chiefly those which concern the spots in the sun, which being frequently both more in number, and greater in bulk than Mercury or Venus ever

> appeared, ought to be presumed to have stronger operations upon the Earth than either of them, being interposed betwixt the Sun and us, or (as they phrase it) in the Heart of the Sun. (132)

Astrology has not altered its "decrees" in the face of new scientific discoveries, hence its "pretence to Experience [is] Unjust."

Gassendi's invocation of astronomy against astrology was taken further by Henry More, that curious hybrid of Platonism, occultism, and the new science. More also derives much of the content and style of his analysis from Pico, but adds a post-Copernican conception of the so-called "*Station* and *Retrogradation* of the Planets": "they that understand the right *Systeme* of the World, know very well that those *Phaenomena* are not real but seeming: which is a Scurvy Slur to the Astrologers" (More 1680, 57). So the whole notion of "*Dignities* and Debilities" deduced from the planets' apparent movements was based on "a mistake of the System of the World, and ignorance of the Earths annual motion," together with "an Idiotick application of accidents or phrases amongst men" — human categories of hierarchy — and was merely "imaginary" (67). Astrology is irremediably linked to Greek cosmological ideas, but there is "no such *Zodiack* in Heaven," or rather no heaven for the fixed and artificial zodiac it constructs, replete with trigons and "triplicities":

> For this Heaven, and this *Zodiack* we speak of is only an old error of *Ptolemie's* and his followers, who not understanding the true *Systeme* of the World, and the motion of the Earth, in which is salv'd the *Anticipation* of the *Æquinoxes*, have phansied a Heaven above the *Coelum stellatum*, and a *Zodiack* that did not recede from West to East as the Starry Zodiack does. (63)

Like Gassendi, More points out the discrepancy between "modern Astrologers building their Houses on that *Zodiacus immobilis* or *fixus*," when the real "*Zodiacus mobilis* of the starry Heaven" has receded since Ptolemy's day by more than one sign (87, 100–101, 106–107). Like Pico he invokes "the known principles of *Opticks*" to show that "the Moon, the nearer she is to the Sun, the more she is illuminated ... though by reason of her positure the light is less reflected to us" (82–83). More invites his opponent to "read *Galilaeus* his *Systema Cosmicum*, or *Des-Cartes* his *Principia*" (76), refers him to Gilbert and Descartes on the load-stone (80, 83f.), but warns him that if he knew more it would destroy his "*Divine Science* of *Astrology*" (84). Confident of the consensus of scientific opinion behind him, More states that "there is no Philosopher of note nowadays, but thinks the Planets as inert and opaque Masses of matter as our Earth is" (95, 61), and challenges the astrologers further with the information that "in Venus the Philosophers observe such Phases as are observed in the Moon" (96). Although no one could claim Henry More as a wholly rational empiricist,[61] he can justly state that "I have fundamentally confuted the *whole Art of Astrology*" (74), following Gassendi's lead by adding to Pico's destructive analysis further damaging material from contemporary astronomy.

The strategy of Gassendi and More, invoking the findings of the new sciences to subvert the occult's pretence to scientific status, was one found frequently in the sixteenth and seventeenth centuries. Although the proponents of the "Yates thesis" sometimes write as if the leaders of the scientific Renaissance were all occultists, or never attacked the occult, and that all its opponents were clergymen or other political and social reactionaries, the fact is that the occult was criticized by a wide spectrum of writers, from those actively involved in the invention and propagation of science — Galileo, Kepler, Mersenne, Gassendi, Bacon, Descartes, Boyle — to others with more leaning towards the occult: Van Helmont, Sennert, Walter Charleton.[62] In many of these attacks the basis for rejection is the appeal to experience and observation. The rise of scientific botany, in which, as Agnes Arber noted, the fact that (for the great illustrated herbals of the sixteenth century) plants were drawn afresh meant that for a while the picture was taken from nature while the text, from Pliny or Dioscorides, contradicting visible details, led in time to a wholly fresh start (see Arber 1938, 1953; Reeds 1976). Plants were now valued for themselves; botany began to exist as a subject in its own right, not just as an extension of medicine.

One result of this new investigation of plants was a sceptical attitude to the doctrine of signatures, which was being energetically disseminated at that time by Agrippa, Porta, Croll, and others. The observation of nature at first hand was bound to create conflicts with the preformed planetary or zodiacal schemes of the occult tradition. Arber finds it "satisfactory" that "the theory of signatures was repudiated by the best of the sixteenth-century herbalists." Dodoens, for instance, wrote in 1583 that "the doctrine of the Signatures of Plants has received the authority of no one ancient writer who is held in any esteem; moreover it is so changeable and uncertain that, as far as a science or learning is concerned, it seems absolutely unworthy of acceptance" (Arber 1938, 225). In 1628 Guy de la Brosse criticized the absence of any criteria in the occult to delimit the working of analogy, since one could "imagine any resemblance between a plant and an animal that happened to be convenient. 'It's like clouds, which can be made to look like anything that fantasy can project — a crane, a frog, a man, an army, and other similar visions'." (ibid.)[63]

Although astrological botany continued to flourish, seventeenth-century scientific botanists pursued the question of signatures with some vigor. John Ray, the greatest botanist before Linnaeus, in his pioneering attempt in 1660 to record the plants of Cambridgeshire, devoted a long paragraph to attacking "the foolishness of the alchemists who chatter and boast so loudly of the signatures of plants." This refutation is worth reprinting complete:

> We have paid close attention to the matter and are moved to assert that the signatures are not indications of natural qualities and powers impressed on plants by nature. (1) Of the plants specifically said to be appropriate to a particular part of the body or to a disease far the greatest number have no signature, as it is easy to demonstrate in the heart, the chest, the skull and the liver. (2) Different parts of the same plant such as the leaves, roots, flowers and seeds, have signatures not merely different but contradictory. (3) Many plants resemble natural or artificial objects for which they have no affinity, as Orchid flowers look like spiders, frogs, bees or butterflies. Between these (as I may note in passing) so great a similarity exists that no one could fail to recognize it immediately nor could he persuade himself that it had ever come about by chance; it is clear proof of intention on the part of nature. The seed cases of Shepherd's purse resemble little bags, Thlaspi small trays, Antirrhinum vulgaris a calf's head; Pisum cordatum has the shape of a heart impressed on it as have the leaves of some species of Medick; Tragopogon a goat's beard; the root of the earth nut a mouse, Iris and Gladiolus a sword, the foliage of fenugreek horns: to which objects they cannot be related. (4) Parts of some plants represent parts of the body with which they violently disagree. Thus the fruit of Anacarditum represents a heart and is nevertheless poisonous; the juice of the spurges is like milk but no one is such an imbecile as to give it to nursing mothers; the flesh of Mespilus is like human excrement and similar in colour, but is not suitable as an aperient. The fungus ignarius growing on the trunks of trees recalls the lungs in shape but is nonetheless harmful and dangerous to them. It would be endless to enumerate every example. (5) The same parts of the body are affected by different and often contradictory diseases which require remedies differing in character. (6) In different plants, parts of the same appearance and shape possess different and diverse properties; the bulbous roots of some, like Narcissus, promote vomiting while those like the root of tulip form a very pleasant food. (7) Neither are the number of signatures so great nor the signatures they bear so obvious

> and plain to everybody that they suggest a pointer or deliberate plan on the part of Nature. There is such a vast number of plants that even if they had come into existence altogether and by chance, any ingenious and imaginative person could have found as many signatures as are known today. (Ray [1660] 1975, 110)[64]

That can stand as a model in applying both Pico's categories of "reason" and "experience." Ray is especially good at picking out "resemblances" that have no "affinity" to the object to which they would be related by superficial similarities.

Ray was one of the leading lights of the Royal Society; their counterpart in Paris, the Académie des Sciences, reached an equally negative conclusion, as the indefatigable Denis Dodart recorded in 1679. Discussing "the virtues of plants"[65] he notes that knowledge of these could be increased in three ways: by avoiding the faults scientists have made in observing and reporting these qualities, by using experience to confirm or refute reports, and by outlining new ways of gaining knowledge, "whether by experiment or by reasoning."[66] Previous writers have confused "the true, the false, the superstitious,"[67] but by working on more stringent lines his "Compagnie" intends to establish truth, for descriptions only have merit when based on experience or delivered by a credible source. On both counts the doctrine of signatures delivered in Porta's *Phytognomica* is found wanting:

> the company ... would like to be able to establish signs of the virtues that would be true and reliable. It has by no means found others in which one can have any confidence, except for changes of color and consistency, which will be explained below, and which only indicate savours. It does not despair of being able to present in the project it has begun material that relates directly to the virtues — but it cannot promise this. The signatures that Jean-Baptiste Porta includes under the title "physiognomy of plants" are purely signs; one can say that nothing could be expected from these signs (at least with regard to use, which has vastly different consequences than speculation) were they not at least as doubtful as they seem to be true to those who would make themselves famous by this opinion. Intelligent persons, who want to reflect on the foundations of this art, and know enough about plants to recognize that the consequences one draws from the facts on which it rests in the main are destroyed by other more accurate and more numerous facts, will not accuse us of having neglected anything that can be of assistance in knowing the virtues of plants by leaving the cultivation of this art to others who are more convinced by it than we are. (Dodart 1679, 66–68)[68]

The note of scepticism there, while refuting the whole basis of astrological botany — "they are purely signs", with no reference to reality — impresses by the coolness and confidence with which it invokes "other facts, more precise and more numerous," that destroy the theory of signatures. Yet although we might expect "rationalism" or scepticism in the 1670s, the point on which I should like to insist is that these qualities are not the result of the revival of ancient scepticism or the influence of Descartes, but are common to the tradition of Renaissance humanism going back to Pico.

It is significant that some of the critics of this tradition were not botanists but doctors, who resented the continuing domination of astrological medicine. Jean Forget, "Medicus Lotharingus," wrote a book attacking Porta, Croll, and others (Forget 1633), while a Belgian doctor did the same (Spiegel 1606; cf. De Virville 1957, 118). Increasingly we find the awareness that astrology and other occult sciences were built on analogy and similitude; as Henry More said of his astrologer opponent, "To shuffle off this invincible Argument he abounds with similitudes, as if he were unacquainted with that trite Aphorism, *Similia non probant*" (More 1680, 111). Daniel Sennert, a Galenist physician but nonetheless an exponent of chemical medicine, attacked Paracelsus for basing his system on analogies taken literally, such as the macrocosm-microcosm, or building retorts to distil urine in the size and shape of a man (Sennert 1662, 121). It is "foolish" to rely on such resources, for "the soul that loves truth is not satisfied with similitudes only, but desires solid demonstrations; and evolves things from their own, not from the principles of another." The "Analogy between the great and little World" has no status as scientific proof: "It is foolish, without sense or experience to flie to such Analogical proofs" (ibid., 28, 96).

Another vehement critic of Paracelsus was John Baptist van Helmont, who rejected the concept of "Microcosm or little world" as "Poetical, Heathenish, and metaphorical, but not natural, or true." It was "phantastical" of Paracelsus to claim that all the elements in the universe are found in man, and are to be applied to "the art of healing: But the life of man is too serious, and also the medicine thereof, that they should play their own part of a Parable or Similitude, and metaphor with us" (Van Helmont 1662, 323). In outlining "A Modern Pharmocopolium and Dispensatory," Van Helmont rejects the whole doctrine of signatures, the basis of astrological medicine:

> I believe that God doth give the knowledge of Simples to whom he will, from a supernatural grace: but not by the signes of nature! For what palmestrical affinity hath the Boars tooth, the Goats blood, the peisle of a Bull, the dung of a Horse, or the Herbe Daysie, with a Pleurisie? or what signatures have those Simples with each other? (Ibid., 458)

Rejecting these traditional remedies, Van Helmont also rejects the subordination of the Earth to planetary influences, affirming that organic growth is a self-contained process. The Earth itself has a "seminal virtue," and "the whole property of Herbes is from their seed," not from the lights of heaven. Traditional astrological botany correlated lists of plants with the seven planets or twelve zodiacal signs, but how could "so few Stars contain the essences, seeds, faces, and properties perhaps of five hundred plants, differing in their species and internal properties?" Like Pico and so many critics of the occult, Van Helmont typically resorts to the flat denial: it is not so.

> Away with these trifles! The properties of Herbes are in the Seeds, but not in the Heaven or Stars. The powers of the Stars are grown out of date, the which by an old Fable have stood feigned unto heats, colds, and complexions. (ibid., 458)

The whole interrelation of astrology, botany, and medicine is but "an old fable." Kepler practised astrology, and tried to develop its scientific potential by linking it with geometry and musical harmonics,[69] but on this side of it he would have agreed with Van Helmont. He, too, rejected astrological *melothesia* and its application to phlebotomy:

> I do not hereby wish to have defended those fantasies ... about assigning a person's limbs to the twelve [zodiacal] signs, scheduling blood-letting according to these assignments, dividing the twelve signs among the planets and the recurring signs. For these childish observations have nothing in common with my thinking. (Kepler 1610)

The common element in so many of the Renaissance rejections of the occult is the denial of what I have called "category-fit," the correlation of two or more categories with the assumption that the resulting grid can be read horizontally, and attributes of the dominant matrix transferred across the parallel columns. The correlation of the planets with the metals, a basic idea in alchemy, was also rejected by several leading scientists in the seventeenth century. William Gilbert, in *De Magnete* (1600), dismissed the idea that metals exist because of some "specific virtues of matter, as the alchemists fondly imagine," claiming that they come into existence when the conformation of the Earth coincides with "fit matter," or that "those metals take from universal nature the forms by which they are perfected." If this were true, the kinds of metals would be vague and undefined, whereas

> In fact the varieties are very few, hardly ten in number. But why nature

> should be so grudging in the number of metals, or why there should be even so many metals as are recognized by man, were not easy to explain, though simpletons and raving astrologers refer to the several planets their respective metals. But neither do the planets agree with the metals nor the metals with the planets, either in number or in properties. For what is common between Mars and iron, save that, like many other implements, swords and artillery are made of iron? What has copper to do with Venus? Or how does tin, or zinc, relate to Jupiter? These were better dedicated to Venus. But a truce to old wives' talk. (Gilbert 1600, 36)

Some 70 years later, in his *Cours de Chymie*, Nicolas Lemery made a dispassionate analysis and rejection of the occult's analogical interrelating of categories:

> *Astrologers* have conceited that there was so great an affinity and correspondence between the Seven *Metals* before named, and the seven *Planets*, that nothing hapned to the one, but the others shared in it ...
>
> For these reasons they have given these seven *Metals* the name of the seven *Planets*, each accordingly as they are governed: and so have called *Gold* the *Sun*, *Silver* the *Moon, Iron Mars, Quicksilver Mercury, Tinn Jupiter, Copper Venus*, and *Lead Saturn* ...
>
> Again they have taught us that the seven *Planets* do every one govern some particular principal part of our bodies; and because the *Metals* do represent the *Planets*, they must needs be mighty *specifick* in curing the distempers of those parts, and keeping them in good plight. Thus they have assigned the *Heart* to *Gold*, the *Head* to *Silver*, the *Liver* to *Iron*, the *Lungs* to *Tinn*, the *Reins* to *Copper*, and the *Spleen* to *Lead.*
>
> Thus you see in short what some of the most *sober Astrologers* do fancy concerning *Metals*, and they draw consequences from hence, which 'twould be too long here to relate. I have told you what the soberest among them say; for nothing can be so absurd as what some of them would have us believe.
>
> 'Tis no hard matter to disprove these conceits, and shew how groundless they are; for no body ever yet got near enough to the *Planets*, to satisfie himself whether they are really of the same nature with *Metals*, or whither any *Effluviums* of bodies do fall from them to us.
>
> Nevertheless if we could be satisfied that Experience did confirm what these persons have offered to maintain, we might then have some reason to think there were some likelihood in their doctrine, although

> their principles were found to be altogether false; but in truth there's nothing to confirm their Opinion, and we find it every day plain enough, that the Faculties and Virtues are utterly false, which they do attribute to the *Planets* and *Metals*; the *Metals* indeed are of good use in *Physick*, and excellent *Remedies* may be drawn from them; but their effects may better be explicated by Causes nearer at hand than the *Stars*. (Lemery 1686, 46–48)

Pico's appeal to "everyday experience," Lemery's "every day plain enough": in this two hundred year span astrology, like the other occult sciences, was shown to be a man-made construct, made, that is, out of human preoccupations and designed to satisfy human desires. Analogies, symbols, similitudes came to be seen as what they really are, rhetorical devices, not logical proofs. In the end astrology was shown to be an art of language.

Notes

1. This version of Garin's *Lo Zodiaco della Vita* (Bari 1976) has a good claim to be the worst translation ever published of a scholarly work, being in many places unintelligible without reference to the original text. See the review by Michael J.B. Allen in *Renaissance Quarterly*, 36 (1983):577–80.
2. For further discussion of the Yates thesis and its critics, see my "Introduction" to Vickers 1984a, 1–55, at 3–9 and notes 9, 10, 14.
3. I have based this analysis of Greek astrology on two of the leading authorities: Bouché-Leclercq 1899; (references abbreviated in the text as "B-L"); and Gundel 1972 (references abbreviated in the text as "Z"). I have also used Boll 1894; Thiele 1898; Boll, Bezold and Gundel 1926; Cumont 1912; Gundel 1936a, 1936b; Gundel 1968; Festugière 1944; and Bidez and Cumont 1938. The primary sources that I have used are: Ptolemy, *Tetrabiblos*; and Firmicus Maternus, *Matheseos Libri VIII*. On critiques of astrology in antiquity see Amand 1945, a classic study, and Long 1982.
4. "cette proposition, pour les adeptes de la philosophie mystique issue de Pythagore et de Platon, n'avait pas besoin d'être démontrée."
5. "des dieux, et même des dieux créateurs des organismes terrestres."
6. "peut causer des méprises, étant le nom de la planète Jupiter."
7. "De là une série d'associations d'idées incohérentes et divergentes, dans lesquelles les astrologues ont choisi à leur gré de quoi composer le type de l'astre puissant et redouté, celui qui trône au plus haut du ciel et pèse irrésistiblement sur les leviers célestes."
8. "foule d'analogies et d'arrangements," "toutes ces idées incohérentes."
9. "le type astrologique de Saturne."
10. "ressemble à des bonds rapides succédant à de courtes rétrogradations."
11. In addition to the works listed in note 3 see Van der Waerden, 1954.
12. Cit. Gundel, *Zodiakos* 481 from Geminus, *Eisagoge*, I:3f; my translation.
13. "une violence faite à la nature."
14. "une mine inépuisable de combinaisons."
15. "n'ont fait que limiter le nombre des combinaisons calculables et disqualifier certains types."

16. "combinaison hybride."
17. "précision minutieuse."
18. "jeu de patience."
19. "sectionnements que les astrologues multipliaient à plaisir pour incruster sur tous les points de la sphère des influences diverses entre lesquelles ils pussent choisir. Il y avait des combinaisons de toute sorte, consistant en répartitions soit des planètes dans les signes,... soit même des signes dans les signes, diversement ordonnés, tantôt suivant l'ordre "naturel" des planètes ou des signes, tantôt suivant l'ordre des trigones, et découpant les signes en fractions de plus en plus petites, tiers, septièmes, neuvièmes, douzièmes, ou même trentièmes, c'est-à-dire aboutissant au degré simple."
20. "combinaison," "correspondances," "distribution," "répartition," "partager," "distribuer," "attribuer," "associer" and "attacher." See, e.g., B-L, "combinaison": 82, 100, 157, 218; "correspondance": 76, 92n, 106, 110, 149, 160, 217, 228, 242, ...; "attribuer"/ "attacher": 94, 97, 98, 99, 101, 103, 111, 112, 131, 132 and n., 154, 158, 162, 165, 167, 172, 181, 183, 185, 192, 200, 215, 230, 248, ...; "répartition": 182, 206, 213, 215n, 224, 227, 230, 246.... Bouché-Leclercq never conceived of this development of classification to be an end in itself as a general principle of astrological methods, but he provided all the evidence necessary for a formal analysis in these terms.
21. "n'était pas encore assez compliqué pour les astrologues en quête de raisons et surtout d'échappatoires."
22. "L'imagination des astrologues est inépuisable, et qui les fréquente doit s'armer de patience."
23. "un système fabriqué par de véritables assembleurs de nuages."
24. "n'est pas, tant s'en faut, un chef-d'œuvre: mais il suffit pour nous donner une idée de ce qu'un virtuose eût pu faire, avec les mêmes pièces sur l'échiquier, en introduisant dans l'analyse ..."
25. "Fécondée par l'intérêt personnel et le loisir, leur imagination dut produire divers systèmes de chronocratories régissant les années, mois, décades, jours et heures: les uns réguliers et relativement intelligibles, les autres irréguliers, inintelligibles et d'autant plus prisés comme arcanes révélés."
26. "Le nombre des arrangements possibles devient vertigineux."
27. "combinaison baroque."
28. "dans l'embarras et nous évader enfin de cette géhenne arithmétique."
29. Bouché-Lelercq notes that because the Pythagorean system, alternating male and female with even and uneven, was used, "that is perhaps why Aries replaced Cancer as the first sign, since one can hardly assign the masculine sex to Cancer without taking it away from Leo and giving it instead to Virgo — and even for astrologers that would be absurd." ("C'est peut-être là la raison qui a fait substituer le Bélier [Aries] au Cancer comme signe initial, car on ne pouvait attribuer le sexe masculin au Cancer sans l'enlever au Lion et l'attribuer par contre à la Vierge, ce qui eût paru absurde même à des astrologues.")
30. "cette science mystico-mathématique."
31. "les signes étaient des êtres vivants."
32. "la figure efficace et bienfaisante par excellence."
33. "antipathique et défavorable."
34. "La théorie des aspects, la création la plus admirée de l'astrologie savante et la base de tous ses calculs, repose en dernière analyse sur les propriétés occultes adjugées aux nombres et aux figures par les Pythagoriciens, propriétés que ressassent *ad nauseam* tous les mystiques et tous les deféenseurs de l'astrologie."
35. "étaient des gens fort ingénieux, et ils n'étaient pas embarrassés de justifier un arrangement qui leur était visiblement imposé par la tyrannie des chiffres."
36. "L'extravagance consiste à confondre la présence réelle d'un astre avec son domicile,

et à supposer que la planète, se déplaçant d'un domicile à l'autre, se "conjoint" par aspect diamétral, trigone, quadrant, avec le Soleil ou la Lune *supposés immobiles.*"

37. "les Poissons qui, eux, n'ont pas de pieds, mais compensent cette fâcheuse inaptitude par le fait qu'ils sont deux."
38. See Dick 1946, 431: "From about 1700 onward the pseudo-science dwindled swiftly, and no further texts appeared for seventy-five years"; and Chapman 1979, 300: "By the end of the seventeenth century the explication of human afflictions in terms of celestial influences had ceased to be a serious part of medicine."
39. Fludd applied his "weather-glass" to urine analysis, but derived the colour symbolism from a fifteenth-century manuscript now in the Bodleian.
40. See, e.g., essays by Paola Zambelli and Ottavia Niccoli on the predicted universal flood of 1524 in Carfagnini 1982, 291–392, and Vickers 1984b.
41. All references are to the edition by Eugenio Garin, with facing Italian translation, 2 vols. (Florence, 1946, 1952), incorporated into the text. Apart from Garin's introduction, the most useful studies of Pico's astrology are Shumaker 1972, 16–27, and Rossi 1965. See also North 1980, 203ff.; Fitzgerald 1969; and Walker 1958, 54–59, who brings out Pico's debt to Ficino.
42. See, e.g., I:36, 38, *78*, 82, 96, 112, 126f., 160, 164, 174, 180, 426, *444*, 604, 610; II:12, 284, 484, 502, 520, 522, 528. Italicized numbers refer to particularly striking formulations.
43. See, e.g., I:*134*, *138*, 442f., *446*, 474, 478f., *480*, *482*, 484, 486f., 492, 520, 546; II:274, 498, 504, 522. Italics for emphasis. On Pico's affirmation of human dignity see his "Oration on the Dignity of Man" (Pico [1948]). See also Garin 1983, 90ff.; Garin 1937, 186–90; and Baron 1927, 145ff.
44. See, e.g., I:*44*, *46*, 80, 100, *110*, 112, 116, 292, 298, *362*, 444, 466; II:378, 464, 522. Pico is here drawing on the tradition of the *vita activa*, with its call for all human activity to be directed to matters of "use and benefit to mankind": see Garin 1965; Rice 1968; and Vickers 1984c, 281–313.
45. See, e.g., I:40, *50*, *58*, 110, 164, 170, 174, 604, 612; II: 206, 502, 504.
46. See, e.g., I:*262*, 264, *278*, 360, 364, 380, 384, 402, 464; II:42, 44, 440.
47. See, e.g., I:190, 192, 194, 210, 274, 390, 392, 492; II:76, 222, 466, 500.
48. See, e.g., II:18, 50, 66, 106, 108, *356*.
49. See, e.g., I:280, *348*, 354, *374*, 376, *466*, 484; II:*24*, 28, 440, 442, 448, 458, 460, 466.
50. See I:160, 162, 164, 166, 168, 190.
51. See II:458, 460, 462, 464, 468, 472, and Schmitt 1969, 85–92.
52. See, e.g., I:86, 100f., *108*, 116, 140, 170, 178, 180, *190*, 210, *262*, 298; II:8, 28, 72, 116, 130, 520.
53. See, e.g., I:54, 136ff. (the chapter title of II, vi: "*Fallacem esse hanc artem, si qua sit, tum propter plura, tum propter discrepantes astrologorum sententias*"), 172, 188, 350, 366, 472, 566 (chapter title: "*Quis ordo servandus, et Alliacensem aliud alias sensisse*"), 570; II:10 (chapter title of VI, iii: "*...quam in earum viribus explicandis astrologi inter se pugnant*"), 28, 44, 60, 104, 116, 124, 126f., 134, 146, 180, 182 (chapter title of VII, vi: "*Non posse initium unum constitui, de quo totius anni pendeant fata; tum varia a variis anni statui principia...*"), 184, 200ff., 208, 214, 226f., 236, 242, 248, 256, 264, 286, 314f., 326 (chapter title of IX, ix: "*Variae sententiae de planetarum motibus, anni inaequalitate, ex maxima Solis declinatione*"), 338 (chapter title of IX, x: "*Dissentes sententiae ...*", 372, 392, 414f., 416 (chapter title of IX, xi: "*Quod cui membro signum dominetur varie a variis determinari, et ab omnibus parum rationabiliter*"), 428, 434, 436, 446, 484, 514.
54. See, e.g., I:86, 96, 98, 100, 106 (chapter heading), 140, 176f., 178, 246, 264, 276, 370, 384, 530, 546; II:8, 298, 304, 360 (their arguments are not even as strong as the rhetoricians' enthymeme — a weaker form of the syllogism), 368, 386 (Avenazra has

used "cachinnis, potius quam syllogismis": cackles), 402, 446, 454 ("nihil sequens antecedentibus cum sequentibus"), 466.

55. See, e.g., I:522, 524; II:130 ("Vanissima igitur dogmata astrologorum, quae nec rationibus firmant, nec experimentis, quando in illis nugantur, in istis non concordant"), 436, 456.
56. See, e.g., I:260, 278, 280, 292, 294; II:20, 38, 40, 128, 222, 246, 424.
57. See, e.g., for the term "fables," I:78, 110, 360, 496; II:30, 274, etc.; for "fictions," see e.g., I:170, 370, 530; II:8, 36, 206, 222, 254, 286.
58. See, e.g., I:296, 544, 612; II:6, 174, 190, 192, 274, 302, 388f., 444, 448.
59. The works I have read — more on the lines of a representative sampling than an attempt at completeness — are: Howard 1583; Fulke 1560; Perkins 1585 (Perkins was shown to be the author by Dick 1939), 1592, 1604; Chamber 1601; Geree 1646, the source for Allen 1659; Rowland 1652; Gassendi 1659; More 1680.
60. See Bauckham 1975: a useful paper, but entirely unaware of Fulke's debt to Pico; and Kocher 1953, 211–12.
61. More blends his Cartesianism with Neoplatonism (e.g., 55), speculates that other suns might have planets inhabited by "Genii or Angels ... having Bodies of Aether or the Celestial matter" (76); he accepts the efficacy of the weapon-salve (78–79, 126), and believes in "Critical days" in medicine (93).
62. For discussion of these attacks, together with further documentation, see Vickers 1984a, the essays by W. L. Hine, R. S. Westman, Ian Maclean, Edward Rosen, Judith V. Field and myself.
63. "C'est comme des nuées que l'on fait ressembler à tout ce que la fantaisie se représente, à une Grue, à une Grenouille, à un Homme, à une armée, et autres semblables visions."
64. On Ray's life-long opposition to the occult see also Raven 1942, 9, 10, 50, 70, 98–99, 102, 159, 276, 309, 380, 430, 453, 464.
65. "Des Vertus des Plantes," Dodart 1679, 45ff.
66. "soit par l'expérience, soit par le raisonnement."
67. "le vray, le faux, le superstitieux."
68. "la Compagnie ... désireroit pouvoir establir des signes des vertus qui fussent véritables & fidèles. Elle n'en a point trouvé d'autres ausquels on puisse prendre quelque confiance, que les changemens de couleur & de consistence, qui seront expliquez dans la suite, & qui ne marquent que les saveurs. Elle ne désespère pas qu'il ne s'en puisse présenter dans le travail qu'elle a commencé, qui se rapporteront directement aux vertus, mais elle ne le peut promettre. Les signatures que Jean-Baptiste Porta comprend sous le nom de Physiognomie des Plantes, sont bien des signes purement tels; & on peut dire qu'il n'y auroit rien à souhaiter dans ces signes (au moins dans ce qui regarde l'usage, qui est bien d'une autre conséquence que la spéculation) s'ils n'estoient au moins aussi douteux qu'ils paroissent véritables à ceux qui se sont voulu rendre célèbres par cette opinion. Les personnes intelligentes, qui voudront faire quelque réflexion sur les fondemens de cet Art, & qui sçauront assez les Plantes pour reconoistre que les conséquences que l'on tire des faits sur lesquels il est principalement establi sont destruites par d'autres faits plus précis, & en plus grand nombre, ne nous accuseront pas d'avoir negligé rien d'avantage à la conoissance des vertus des Plantes, en laissant le soin de cultiver cet Art à d'autres qui en seront plus persuadez que nous ne sommes...."
69. The best evaluation of Kepler's astrology is by Simon 1979. See also Hammer 1971; Aiton 1976; and Field 1984. Graubard 1968 has some valuable material but also some confusion.

References

I. Ancient and Renaissance Sources

Allen, J., 1659. *Judicial Astrologers Totally Routed, and their Pretence to Scripture, Reason & Experience, Briefly, yet Clearly and Fully Answered*. London.

Calvin, John, 1560. *An Admonicion against Astrology Iudiciall*, trans. Goddred Gilby, London; new "Edition critique," *Avertissement contre l'Astrologie Judiciaire*, ed. O. Millet. Geneva, 1985.

Cardanus, H., 1658. *Metoposcopia libris tredecim*. Paris.

Chamber, John, 1601. *A Treatise Against Judicial Astrologie*, together with *Astronomiae Encomium*, in Latin and English. London.

Croll, Oswald, 1609. *Tractatus de signaturis internis rerum* ..., Frankfurt; trans. in *Royal and Practical Chymistry*. London, 1670.

Dodart, Denis, 1679. *Mémoires pour servir à l'histoire des Plantes*, Seconde Edition. Paris.

Firmicus Maternus, 1975. *Matheseos Libri VIII*, trans. J.R. Bram as *Ancient Astrology. Theory and Practice*. Park Ridge, N.J., 1975.

Forget, Jean, 1633. *Artis Signatae Designata Fallacia*. Nancy.

Fulke, William, 1560. *Antiprognosticon, that is to saye, an Invective agaynst the vayne and unprofitable predictions of the Astrologians*. London.

Gassendi, Pierre, 1659. *The Vanity of Judiciary Astrology. Or Divination by the Stars*. London; from Gassendi's *Opera omnia*. Paris, 1658.

Geminus, *Eisagoge*, ed. and trans. G. Manitius. Leipzig: Teubner, 1898.

Geree, John, 1646. *Astrologo-Mastix, Or A Discovery of the vanity and iniquity of Judiciall Astrologie*. London.

Gilbert, William, [1600] 1893. *De Magnete*, trans. P. Fleury Mottelay. London; repr. New York, 1958.

Howard, Henry, 1583. *A Defensative agaynst the poyson of supposed Prophecies*.... London; rev. ed., 1620.

Kepler, Johannes, 1610. *Tertius Interveniens*, cit. and trans. E. Rosen, in Vickers 1984a, 258.

Lemery, Nicolas, 1686. *A Course of Chymistry*, trans. from the fifth French edition by W. Harris. London.

Manilius, 1980. *Astronomica*, ed. C. P. Goold. London: Loeb Classical Library, 1980.

More, Henry, 1680. *Tetractys Anti-Astrologica, Or, The Four Chapters in the Explanation of the Grand Mystery of Godliness, Which contain a brief but solid Confutation of Judiciary Astrology*.... London.

Perkins, William, 1585. *Foure Great Lyers*. London.
—, 1592. *A Golden Chaine*, in: *Workes*, Vol. 1. London, 1612, 42–44.
—, 1604. *A Commentarie [on] Galatians*, in *Workes*, Vol. 2. London, 1617, 285–87.
Pico della Mirandola, Giovanni, 1496. *Disputationes adversus astrologiam divinatricem*, Bologna; ed. E. Garin, I–II. Florence, 1946, 1952.
—, [1948]. "Oration on the Dignity of Man," trans. E. L. Forbes, in: *The Renaissance Philosophy of Man*, ed. E. Cassirer, P.O. Kristeller and J. H. Randall, Jr. Chicago, 215–54.
Porta, Joannes Baptista, 1588. *Phytognomonica octo libris contenta*. Naples.
Ptolemy, 1970. *Tetrabiblos*, trans. F. E. Robbins. London: Loeb Classical Library, 1977.
Ray, John, [1660] 1975. *Catalogus Plantarum circa Cantabrigiam nascentium* (Cambridge); ed. and trans. A. H. Ewen and C. T. Prime as *Ray's Flora of Cambridgeshire*. Hitchin.
Rowland, William, 1652. *Judiciall Astrologie Judicially Condemned*. London.
Sennert, Daniel, 1662. *Chymistry Made Easie and Useful. Or, The Agreement and Disagreement of the Chymists and Galenists*, trans. N. Culpeper and A. Cole. London.
Spiegel, A., 1606. *In rem Herbarium Isagoge*.
Swift, Jonathan, 1941. *The Prose Writings of Jonathan Swift*, ed. H. Davis, 14 Vols. (Oxford, 1939–1968), Vol. XI: *Gulliver's Travels*.
Van Helmont, John Baptist, 1662. *Oriatrike or Physick Refined*, trans. J. Chandler, London, *from Ortus Medicinae*. Amsterdam, 1648.

II. Modern Sources

Aiton, E. J., 1976. "Johannes Kepler in the Light of Recent Research," *History of Science* 14:77–100.
Allen, D. C., 1941. *The Star-Crossed Renaissance. The Quarrel about Astrology and its Influence in England*. Chapel Hill, N.C.; repr. New York, 1966.
Larkey, S. V., 1935. "Astrology and Politics in the First Years of Elizabeth's Reign," *Bulletin of the Institute of the History of Medicine* 3:171–186.
Amand, D., 1945. *Fatalisme et Liberté dans l'Antiquité Grecque*. Louvain.
Arber, A., 1938. *Herbals. Their Origin and Evolution*, 2nd ed. Cambridge; repr. Darien, Conn., 1970.
—, 1953. "From Medieval Herbalism to the Birth of Modern Botany," in:

Science, Medicine and History, ed. E. A. Underwood, 2 vols. London, I:317–36.

Baron, H., 1927. "Willensfreiheit und Astrologie bei Marsilio Ficino und Pico della Mirandola," in: *Kultur und Universalgeschichte. Festschrift für W. Goetz*. Leipzig and Berlin, 145ff.

Bauckam, R., 1975. "Science and Religion in the Writings of Dr. William Fulke," *British Journal for the History of Science* 8:17–31.

Bidez, J., and F. Cumont, 1938. *Les Mages Héllénisés. Zoroastre, Ostanes et Hystape d'après la Tradition Grecque*. Paris.

B-L, see Bouché-Leclerq 1899.

Bleker, J., 1976. "Chemiatrische Vorstellungen und Analogiedenken in der Harndiagnostik Leonhard Thurneissers (1571 und 1576)," *Sudhoffs Archiv* 60:66–75.

Boll, F., 1894. *Studien über Claudius Ptolemäus*. Leipzig.

Boll, F., C. Bezold, and W. Gundel, 1926. *Sternglaube und Sterndeutung. Die Geschichte und das Wesen der Astrologie*, 3rd ed. Leipzig; 5th ed., with bibliographical appendix by H. G. Gundel. Darmstadt, 1966.

Bouché-Leclerq, A., 1899. *L'Astrologie Grecque*. Paris; repr. Brussels, 1963.

Chapman, A., 1979. "Astrological Medicine," in: *Health, Medicine and Mortality in the Sixteenth Century*, ed. C. Webster. Cambridge: 275–300.

Clagett, M., 1957. *Greek Science in Antiquity*. London.

Coopland, G. W., 1952. *Nicolas Oresme and the Astrologers*. Liverpool.

Cumont, F., 1912. *Astrology and Religion among the Greeks and Romans*. New York.

De Virville, D., 1957. "De l'influence des idées préconçues sur le progrès de la botanique du XVe au XVIIIe siècle," *Revue de l'Histoire des Sciences* 10:110–19.

Dick, H. G., 1939. "The Authorship of *Four Great Lyers* (1585)," *The Library*, 4th Ser., 19:311–14.

—, 1944. "Intellectual Background: the War against Judicial Astrology", in: Tomas Tonkis, *Albumazar: A Comedy*, ed. Dick. Berkeley and Los Angeles.

—, 1946. "Students of Physic and Astrology. A Survey of Astrological Medicine in the Age of Science," *Journal of the History of Medicine* 1: 300–315, 419–33.

Festugière, A., J., 1944. *La Révélation d'Hermès Trismégiste. I: L'Astrologie et les Sciences Occultes*. Paris.

Field, J. V., 1984. "A Lutheran Astronomer: Johannes Kepler," *Archive for History of Exact Sciences* 31:189–272.

Fitzgerald, D. J., 1969. "Some notes on Pico's Dispute with Astrology," in: *Arts Libéraux et Philosophie au Moyen Age*. Montreal and Paris: 1049–55.

Garfagnini, G., 1982 (ed.). *Scienze Credenze Occulte Livelli di Cultura*. Florence.

Garin, E., 1937. *Giovanni Pico della Mirandola*. Florence.

—, 1965. *Italian Humanism. Philosophy and Civil Life in the Renaissance*, trans. P. Munz. Oxford.

—, 1983. *Astrology in the Renaissance. The Zodiac of Life*, trans. of *Lo Zodiaco della Vita*, Bari, 1976, by C. Jackson, J. Allen, and C. Robertson. London.

Geertz, C., 1956. "Religion as a Cultural System," in *Anthropological Approaches to the Study of Religion*, ed. M. Banton. London: 1–46.

Godwin, J., 1979. *Robert Fludd. Hermetic Philosopher and Surveyor of Two Worlds*. London.

Graubard, M., 1968 "Astrology's Demise and its Bearing on the Decline and Death of Beliefs," *Osiris* 13:210–61.

Gundel, H. G., 1968. *Weltbild und Astrologie in den griechischen Zauberpapyri*. Munich.

—, 1972. *Zodiakos. Der Tierkreis in der antiken Literatur und Kunst*, Munich, with section on the Zodiac in the ancient orient by Robert Böker; a separate issue of Pauly-Wissowa, *Realencyclopädie der klassischen Altertumswissenschaft*, Band XA, cols. 461–710.

Gundel, W., 1936a. *Dekane und Dekansternbilder*. Glückstadt and Hamburg.

—, 1936b. *Neue astrologische Texte des Hermes Trismegistos*. Munich.

Hammer, F., 1971. "Die Astrologie des Johannes Kepler," *Sudhoffs Archiv* 55:113–35.

Jüttner, G., 1971. "Die Signature in der Pflanzenabbildung." *Pharmazeutische Zeitung* 51:1998–2001.

—, 1978. "Alchemie und Signaturenlehre in der Therapie der frühen Neuzeit," in: *Der Übergang zur Neuzeit und die Wirkung von Traditionen*. Göttingen, 130–140.

King, L. S., 1973. "Metoposcopy and Kindred Arts," *Journal of the American Medical Association* 224:42–46.

Kocher, P. H., 1953. *Science and Religion in Elizabethan England*. San Marino, Ca.; New York, 1969.

Lauer, H. H., 1966. "Zahl und Medizin," *Janus* 53:161–93.

Larkey, S. V., 1935. "Astrology and Politics in the First Years of Elizabeth's Regine," *Bulletin of the Institute of the History of Medicine* 3:171–86.

Leclercq, H., 1918. "La Médecine des Signatures magiques," *Janus* 23: 5–28.

Lemay, R., 1962. *Abū Ma'shar and Latin Aristotelianism in the Twelfth Century*. Beirut.

Lloyd, G. E. R., 1961. *Polarity and Analogy. Two Types of Argumentation in Early Greek Thought*. Cambridge.

—, 1970. *Early Greek Science, Thales to Aristotle*. London.

—, 1973. *Greek Science after Aristotle*. London.

—, 1979. *Magic, Reason and Experience. Studies in the Origin and Development of Greek Science*. Cambridge.

—, 1983. *Science, Folklore and Ideology. Studies in the Life Sciences in Ancient Greece*. Cambridge.

Long, A. A., 1982. "Astrology: Arguments Pro and Contra," in: *Science and Speculation. Studies in Hellenistic Theory and Practice*, ed. J. Barnes, J. Brunschwig, M. Burnyeat and M. Schofield. Cambridge, 165–92.

Madkour, I., 1969. "Astrologie en terre d'Islam," in: *Arts Libéraux et Philosophie au Moyen Age*. Montreal and Paris, 1041–47.

Müller-Jahnke, W.-D., 1982. "Zum Prioritätenstreit um die Metoposkopie: Hàjek contra Cardano," *Sudhoffs Archiv* 66:79–84.

—, 1984. "Ordnung durch 'Signatur'. Analogiedenken und Arzneischatz im 16. und 17. Jarhundert," *Deutsche Apotheker-Zeitung* 124:2184–89.

Neugebauer, O., 1957. *The Exact Sciences in Antiquity*, 2nd ed. Providence, R. I.

North, J. D., 1980. "Astrology and the Fortunes of Churches," *Centaurus* 24:181–211.

Pagel, W., 1958. *Paracelsus. An Introduction to Philosophical Medicine in the Era of the Renaissance*. Basel.

Parr, J., 1953. "Sources of the Renaissance Englishman's Knowledge of Astrology. A Bibliographical Survey and a Bibliography," in: *Tamburlaine's Malady and Other Essays on Astrology in Elizabethan Drama*. Alabama: 112–50.

Pruckner, H., 1933. *Studien zu den astrologischen Schriften des Heinrich von Langenstein*. Leipzig and Berlin.

Quecke, K., 1955. "Die Signaturenlehre im Schrifttum des Paracelsus," *Beiträge zur Geschichte der Pharmazie* 1:41–52.

Raven, C., E., 1942. *John Ray, Naturalist. His Life and Works*. Cambridge.

Reeds, K. M., 1976. "Renaissance Humanism and Botany," *Annals of Science* 33:519–42.

Rice, E. F., 1958. *The Renaissance Idea of Wisdom*. Cambridge, Mass.

Rossi, P., 1965. "Considerazioni sul declino dell'astrologia agli inizi dell'età moderna," in: *L'opera e il pensiero di Giovanni Pico della Mirandola nella storia dell' Umanesimo*, 2 vols., II. Florence, 315–31.

Schmitt, C. B., 1969. "Experience and Experiment: A Comparison of Zabarella's View with Galileo's in *De Motu*," *Studies in the Renaissance* 16:80–138.

Schneider, W., 1950. "Über die Signaturenlehre in Medizin und Chemie," *Die Pharmazie* 5:355–59.

Shumaker, W., 1972. *The Occult Sciences in the Renaissance. A Study in Intellectual Patterns*. Berkeley and Los Angeles.

Simon, G., 1979. *Kepler, astronome astrologue*. Paris.

Thiele, G., 1898. *Antike Himmelsbilder*. Berlin.

Thomas, K., 1973. *Religion and the Decline of Magic*, rev. ed. Harmondsworth.

Thorndike, L., 1923–1958. *A History of Magic and Experimental Science*, 8 vols. New York.

Van der Waerden, B. L., 1954. "History of the Zodiac," *Archiv für Orientforschung* 16:217–30.

—, 1965. *Science Awakening*. Groningen.

Vickers, B., 1973. *Towards Greek Tragedy*. London.

—, 1979. "Frances Yates and the Writing of History," *Journal of Modern History* 51:287–316.

—, 1984a (ed.). *Occult and Scientific Mentalities in the Renaissance*. Cambridge.

—, 1984b. Review of Garfagnini 1982 in: *English Historical Review* 390:114–18.

—, 1984c. "Bacon's So-Called 'Utilitarianism': Sources and Influence," in *Francis Bacon. Terminologia e Fortuna nel XVII Secolo*, ed. M. Fattori. Rome, 281–313.

—, 1988. "On the Function of Analogy in the Occult," *Hermeticism and the Renaissance. Intellectual History and the Occult in Early Modern Europe*, ed. I. Merkel and A. G. Debus. Cranbury, N.J., 265–90.

Walker, D. P., 1958. *Spiritual and Demonic Magic from Ficino to Campanella*. London.

Wedel, T. O., 1920. *The Medieval Attitude towards Astrology*. New Haven.

Yates, F., 1967. "The Hermetic Tradition in Renaissance Science," in *Art, Science, and History in the Renaissance*, ed. C. S. Singleton. Baltimore, 255–74.

—, 1972. *The Rosicrucian Enlightenment*. London.

Z, see Gundel 1972.

Critical Reactions to the Occult
A Comment

RIVKA FELDHAY

It is an honor and a great pleasure for me to comment on Brian Vickers' stimulating paper. First because I have to admit, somewhat to my embarrassment, that the occult itself appeals to my imagination, as it has always done to many naive minds, who have also perhaps felt the need for "a little magic" to cope with the incongruities of reality. Second, more seriously, because the problem of the occult sciences, their revival during the Renaissance, and their decline in the seventeenth century seems to me central to the understanding of the much wider issue of the transition of European culture from traditionalism to modernity. I think that, like Vickers, I am rather struck by the difficulty of reducing this phenomenon to simple terms, and like him I feel a kind of dissatisfaction with recurrent attempts to analyze its different aspects, attempts which, in spite of growing sophistication, have failed to capture it in its totality. In spite of the proliferation of attempts to analyze that transition, many of them are still characterized by a kind of one-sidedness which leaves enough space for further discussion.

As a reaction to previous tendencies to ignore the occult sciences as a significant aspect of the "renaissance" of ancient science in general during the sixteenth century, the occult has become subject to deeper scrutiny in the last twenty years. The attraction of this historiographical novelty has created an alternative tradition, which seeks recognition for the role played by the revival of the occult sciences in the eventual triumph of modernity. Insofar as the occult sciences were socially identified with radical elements who called for reform, closely associated with doubts as to traditional authority in the

E. Ullmann-Margalit (ed.) The Scientific Enterprise, 93–99.

religious sphere, and considered a necessary transitional stage between scholasticism and modern science, their centrality in the process of transition to the modern world seemed to be vindicated. Vickers, if I understand him correctly, feels this interpretation to be no less one-sided than the previous one, which tended, as I have mentioned, to ignore the role of the occult altogether. The need for a modification may follow, in his opinion, from the fact that the revival of the occult sciences was accompanied by simultaneous systematic criticism, which began almost at the same time as the revival itself and assumed the form of an alternative current of thought during the second part of the sixteenth and the first part of the seventeenth century. Vickers emphasizes the need to examine Renaissance arguments *against* the occult, an enterprise which, surprisingly enough, has attracted very little research. He also suggests that this examination be limited to the intellectual aspect, claiming that this approach may show that the decline of the occult was due not only — and perhaps not even mainly — to religious and social factors, as some historians seem to believe, but also to its intellectual "inadequacy."

I could not agree more with Vickers' suggestion that criticism of the occult should be examined as an aspect no less relevant to the understanding of Renaissance culture than the insistence by many historians on the centrality of the occult itself. And although I do not doubt the validity of this enterprise for its own sake, I think it not superfluous to draw attention to its relevance to the process of transition which I have mentioned. For this kind of enterprise seems to me essential to the modification of a major defect in a theory of transition shared — implicitly or explicitly — by both traditions, the historiographical tradition that tended to ignore the occult and the alternative one that emphasized its centrality. Both traditions envisage the change from the traditional world view to the modern world view in terms of linear stages: the first from scholasticism to modern science, the second from scholasticism to the occult and from the occult to modern science. Focusing on criticism of the occult, as Vickers does, one can view the whole process in less simplistic terms.

Vickers' analysis of Pico's critique in terms of Aristotelian criteria of rationality shows that at the end of the fifteenth century Aristotelian modes of thought were by no means barren, completely degenerate, or dead, as a result of either humanistic criticism or the emergence of the naturalistic philosophies of the Renaissance. On the contrary, they were lively, useful, and perhaps still fruitfully developing in response to all of these challenges. The recognition that this is so should bring about a modification of the theory of linear stages. Transition occurred not by stages but through a process of a continuing critical dialogue between different modes of thought, in which

Aristotelianism, humanism, and the occult sciences took part and challenged each other. This complex process of mutual criticism gradually gave birth to the new science.

However, whereas Vickers' innovative definition of an alternative locus of interest seems to lead almost naturally to the recognition of a fruitful critical dialogue, to which the occult had something to contribute, the actual contents of his analysis, both of the occult and of the critical reactions to the occult, in fact lead to a different view. Here there is a genuine disagreement between us, which I will not try to hide or underestimate. I think that this disagreement reveals the openness of this issue at the present stage of research into this fascinating topic. Opinions are varied, and it seems to me right to face this situation directly by presenting, in the fairest possible way, both attitudes and the arguments behind them. I shall try, then, to proceed to the heart of the matter.

Vickers' understanding of the essence of the occult mode of thinking and his reading of its Renaissance critique run as follows. Essentially, the occult sciences are best described as verbal, self-contained, symbolic systems. At their basis lies a set of arbitrary, completely wrong assumptions about the relation between man and nature, according to which there ought to be a correspondence between elements of the cosmos and elements of the little cosmos, namely, man. Specific relations of correspondence can be known to man, since both the macrocosmos and the microcosmos are interspersed with signs ready for interpretation. Understanding and knowledge of the signs — and of correspondences through them — enable one to interfere with the "influences" that operate between the macrocosmos and the microcosmos and vice versa.

On the basis of these assumptions, an imaginary network of specific correspondence relations in the cosmos is built up. Anthropomorphic categories, anchored in an analogical mode of reasoning, are projected onto nature. These categories are subjected to further arbitrary classification, of an ever-growing complication. Deductions are then made from the natural to the human sphere.

The occult sciences, according to this view, are not sciences at all, since they differ from science in their aims as well as in their procedures. No less do they differ from science by their criteria of legitimation. Whereas science is interested in the universal processes of nature for their own sake, the occult sciences are mainly interested in the implications of these phenomena for the fate of the individual human. Whereas science uses categories essential to nature and emanating from it, the categories of the occult are arbitrary and stem from superficial analogies. While scientific explanations are legitimized by experience and logical systematization, the occult way of thinking defies both experience and logic; it both disregards observation and is full of contradictions. The exposition of the occult is summed up by Vickers in the following words:

> The categories of astrology, then, are not natural or physical but cultural, anthropological, anthropomorphic. We are dealing with a sign system and classification system obeying its own increasingly elaborate rules, having totally lost contact with the physical reality that it still pretends to depend on and interpret.

Pico's arguments against astrology, which probably epitomize the later critique as well, do not much differ from the criticism inserted into the exposition of the occult, which I have just summarized. The occult uses anthropomorphic, man-made categories to explain natural phenomena; its predictions, when tested, are wrong; its empirical observations are not obtainable or not accurate; it is confused, full of internal contradictions, and unsystematic.

Vickers' opinion is that this critique was both commonsensical and stood up to universal criteria of rationality ("If the world were wholly rational Pico's refutation of astrology would have killed it outright"). This judgment implies the conclusion that one of the causes behind the eventual decline of the occult was simply the fact that it did not work. Actually, although Vickers does not say so explicitly, the wider implication is that modernity, and especially modern science, developed (in a paraphrase on the words of E. Strong, who came to this conclusion during the 1930s) "in spite of the occult, and not because of it."

But is this really true? I think that there is more than one way to show that it is not, or at least not simply so. At the beginning of his paper, Vickers claims that his aim is "not to summarize astrology as a body of doctrine, rather to draw attention to the mental processes that lie behind it...." In this context it is worth mentioning that all attempts made up to now by anthropologists, sociologists of knowledge, and philosophers of science to analyze the differences between scientific and prescientific, pseudoscientific, or occult systems in terms of "mental processes" have only proved how difficult — almost impossible — it is to analyze these differences by means of such categories as "natural" vs. "arbitrary," "verbal" vs. "real," "consistent" vs. "inconsistent," "systematic" vs. "unsystematic," "rational" vs. "irrational," "causal" vs. "analogical," and so on. On the one hand, comparative anthropologists, especially people like Evans-Pritchard, Horton, and Wynch, have shown that primitive, prescientific symbolic systems do maintain a measure of consistency, systematization, and in this sense rationality, previously almost unsuspected. On the other hand, philosophers of science have pointed out that modern science itself is liable to an unexpected degree of inconsistency, arbitrariness, and defiance of experience. On the basis of this knowledge, one could argue analytically that an analysis of the occult sciences by means of the categories suggested by Vickers is not likely to advance us in our attempt to understand the "mental processes" operating behind them.

However, my historical bias dictates another method. What I would rather do is take a few examples from Pico's most critical comments on astrology and show that, far from reflecting any "universal" rationality, they actually express philosophical presuppositions, deeply anchored in the Aristotelian frame of mind, and no less "arbitrary," "verbal," etc., than the presuppositions of the occult sciences. At the same time, I will try to show that some of the presuppositions of the occult, which superficially seem "arbitrary," "man-made," or "unnatural," actually gave vent to alternative ontological and epistemological premises, which were eventually accepted and developed within a different context by modern science. On the basis of these examples, which are obviously not meant to present anything like an exhaustive analysis, I will then repeat my argument concerning the emergence of the modern conception of reality out of a critical dialogue between Aristotelian rationality, the occult, and new elements which specifically characterize modern science.

Vickers makes much of Pico's demand that a theory meet experimental criteria, thereby showing

> a critical spirit and sustained argumentative power that any modern scholar might envy. "What criterion do astrologers use?" he repeatedly asks. Albumasar is rebuked for accepting everything he read uncritically, neither refuting nor questioning, faults of which Pico was certainly never guilty.

These remarks are rather perplexing. For it is clear that, just as astrologers were uncritical of their own deep convictions, so was Pico uncritical of the most fundamental assumptions of Aristotelian philosophy, which are no less anthropomorphic, from our point of view, than the astrologers'. This fact is well documented by Vickers himself, in the following quotation from Pico. The whole system, says Pico, is "based on a certain similitude with human life, contrary to nature and to the *dignity* of the celestial bodies."

Vickers' logic inspires one to ask whether the "dignity of the celestial bodies" (conceptually implying the Aristotelian belief in the perfection of the heavens versus the corruptibility and mutability of the sublunary region) is not an anthropomorphic concept, or better, whether the "dignity of the celestial bodies" was ever checked against observational evidence. (In the period in question it was common knowledge, based on observation, that, for example, the planets vary in brightness. Did anyone seriously question the perfection of the heavenly bodies as a result of this datum?) One may indeed wonder not only why the occult belief in correspondences was maintained, but also why the main Aristotelian and Ptolemaic tenet of the division between the celestial and sublunary region was maintained, in spite of evidence to the contrary.

Moreover, another of Pico's arguments against correspondences is paraphrased by Vickers as follows: "Some of these category-fits fail because they attempt to correlate things *different by nature.*" This argument echoes the Aristotelian belief in hierarchical essences or natures which, so to speak, fill the cosmic space. However, it is not at all certain that the occult belief in correspondences between "things different by nature" and in influences working through the cosmos was not actually instrumental in throwing away the "false," "arbitrary," and essentially anthropomorphic distinction between various parts of the cosmos, heaven and earth, for example. Indeed, it is more likely the case that, far from diverting men's minds back to anthropomorphism, the assumptions underlying the occult belief in correspondences and influences tended to prepare the climate of opinion necessary for Galileo's telescopic discoveries, which completely discredited the distinction between heaven and earth and supplied the necessary background for the homogeneous concept of space presupposed by Newtonian science.

Pico had a further argument against astrology. It was related to astrology's "dependence on numerical methods, equally man-made and arbitrary. The reciprocal influence of natural bodies depends on natural causes, not on numbers and figures, superimposed categories." This argument, again, is clearly rooted in the Aristotelian belief that mathematical entities, which are essentially intellectual and in any case of a different ontological nature than physical entities, cannot account for physical reality. Needless to say, this assumption is far from representing any canon of universal rationality. Modern science, for one thing, believes the opposite. This is not to say that the occult really showed how mathematical methods could be used to expose natural causality. That part of the modern conception of reality was indeed the unique contribution of modern science. However, was not the occult challenge of the Aristotelian bias against mathematical methods, as well as its insistence on number as a clue to the secrets of nature, instrumental for the emergence of the modern conception of reality?

Contrary to Vickers' belief (and I hope he will correct me if I misinterpret him) in the irrelevance and inadequacy of the occult, I tend to think that the occult did contribute something to the critical dialogue between intellectual currents in the Renaissance. By suggesting alternative ontological and epistemological presuppositions (like the homogeneity of space, or the value of mathematics for the understanding of natural causes) to the predominating Aristotelian views — even if these were only implicit and not actually developed in a systematic manner — it became part of the complex process of mutual criticism and self-criticism, perhaps impossible within the context of a single system of ideas. I definitely identify with Vickers' reaction against the "over-large claims for the occult, elevating it to the status of a dominant or even defining element in Renaissance culture," as he puts it. I agree that more attention should be paid to critical arguments against it. But I doubt the attempt to dismiss it as a relevant element in the development of the modern conception of reality. However, our disagreement lies in questions relating mainly to evaluation and judgment. And on such matters disagreement is perhaps fruitful.

Works of the Imagination

MARTIN WARNKE

When an art historian announces a paper about imagination, perhaps one is inclined to expect painted paradises that project some lofty and harmonious state of being. Today one generally conceives of imagination as the capacity to transcend realities, a power which, once realized, may fulfill all expectations of happiness.

A paper treating this productive and positive capacity would be possible and meaningful. However, the term "imagination" denotes not only a means of compensating for a deficient reality, but also an ability to reveal the authentic existence underlying reality's beautiful but deceitful apppearances. Given this meaning, imagination does not indeed conform to Schiller's definition of it as "the faculty of idealization."[1] It is much broader, also designating the capacity to negate reality by bringing to light things previously suppressed, left unresolved, or obscured by reality.

Coming to terms with this role of imagination in art was already a concern of Cennino Cennini around 1400, when he theorized about fantasia in the earliest treatise on art of modern times, *Il Libro dell'Arte*. Cennini writes:

> Painting calls for imagination as well as skill of hand, in order to discover things that have never been seen, in order to seek beyond the shadow of natural objects and to fix things with the hand, to let that which is not be (quello che non è, sia).... [For as much] as the poet is free to compose and combine or not as he pleases, so is the painter granted freedom to compose a figure, standing or sitting, half man and half horse, as he pleases according to his imagination.[2]

E. Ullmann-Margalit (ed.) The Scientific Enterprise, 101–116.

Thus Cennini demands a liberty for the painter that Horace denies him; that ancient author censures painters for grafting together man and animal in wild configurations, like those produced in delirious dreams (Horace, *Ars Poetica*, 1–13). In contrast, Cennini pleads that the artist be allowed to delve beneath the shadow of natural objects, so that "that which is not may be."

Cennini defines imagination as the license of the painter to combine man with animal, and so to render the impossible, the abnormal and the contradictory both visible and tangible. This astonishing plea has been interpreted as an open recognition of the peripheral area of medieval art that brought forth those fabulous beasts, monsters and chimeras which abound under the eaves of medieval edifices, on the jambs of church portals, on the capitals of columns or in the margins of illuminated manuscripts (Summers 1981, pp. 49ff., 494 n. 93).

Confirmation for this conception may be found in the texts of artists' contracts dating from the fifteenth century. Here one encounters echoes of Cennini's specific formulation, granting the artist the freedom to represent accessory figures, as well as the design of decorative margins, "as he deems best."[3]

In practice, it allowed the artist's imagination free play in corners, on edges and in empty spaces. Filling these, the artist could combine existent and nonexistent, man and animal. Imagination could carry out its task by creating marginal phenomena, phantoms and bizarre curiosities, which do *not* exist in reality.

Proceeding from this point, one might relate the history of art in modern times as the progressive expansion of this peripheral zone, which harbored figures actually repressed or "only" imagined. A history of the imagination would then be also a history of resistance to this expansion, i.e., a history of the attempt to ban imagination to the periphery. That the visual arts were especially affected by such a ban is expressed by Baldassare Castiglione's opinion of Leonardo: "That painter has invented ever new chimeras, which even he, despite all his skill, would never succeed in painting" (Castiglione 1972, Libro II, cap. 39, p. 148). Leonardo's artistic skills would thus have been no match for his imaginative mind. He could not — and was not allowed to — draw or paint the workings of his imagination. The conceivable was not allowed to be possible.

Dürer, also following this line of reasoning, warns the painter against "something impossible and intolerable to Nature," unless he proposes to create some work of feverish dreams that confound everything (Dürer 1969, III, p. 283). Consequently, Dürer himself must have understood his wildly concocted marginalia in the prayer book of the Emperor Maximilian as "a work of dreams." In Italy, such marginal figures established themselves as a separate genre in the School of Raphael. Called "Grotesque," these droll figures and chimeras were tolerated as framing and filling elements, and according to Francisco de Hollanda even Michelangelo might have justified them; for if they appeared false and absurd by any natural measure, they could still be just and fitting for artistic reasons (Summers 1981, pp. 135ff.). Later, Federico Zuccari celebrates these grotesques of the Raphael School as emanations of a "virtu fantastica" (ibid., p. 139), so it seems that Cennini's original demands for artistic liberty were realized and accepted at least in the ornamental borders of late sixteenth-century art.

However, we possess a document from 1573 showing that the guardians of normality had also learned to check in between and along the edges. At that time, Paolo Veronese was hauled before a tribunal of the Inquisition in Venice, because suspicions had been raised concerning several curiosities among the accessory figures of a Last Supper that he had painted for the refectorium of the Santi Giovanni e Paolo monastery. Our testimony of the proceedings is a detailed protocol:

> Paolo, what does it signify that the servant figure has blood streaming from his nose? He answered: I have represented the servant so, because due to some accident blood may have issued from his nose. He was then asked: What do these armed halberdiers in German garb signify? He answered: Regarding this I must utter twenty words. He was informed that he might do this. He said: We painters claim the same license as that taken by poets and fools. I painted both those halberdiers — one who drinks, the other who eats; they stand aside near a blind stairway, ready to perform some service. This seemed appropriate, since the lord of the house, as I understand, was a great and rich gentleman, who might have kept such servants. One asked him: That fool with the parrot perched on his fist — why have you included him in the painting? He answered: My dear sir, for decorative purposes (*per ornamento*) as one generally does.... One asked him: Have you been assigned by anyone to paint Germans, fools, and the like? He replied: My dear sir, no. But the contract says that I should embellish the painting as I see fit (*secondo me parese*). The painting is large enough to contain many figures of my own judgement. One said to him: If the ornaments (*ornamenti*) you painters usually add are demanded, is it not the rule to make them correspond with the subject and main figures, or is it perhaps the rule that you add them as you like without considering or evaluating, and according to the whims of your imagination? He

> answered: I paint pictures attending to whatever appears appropriate to me and to whatever I can grasp with my understanding.[4]

Thus the Inquisition shows itself wary of those peripheral and intermediate zones contractually left to works of the artist's imagination, places where animal and man, existent and nonexistent were grafted together, refuges for negating or subversive activities that had to be controlled.

Veronese's interrogation by the court of the Inquisition demonstrates how the imagination's capacity to discover monsters, chimeras and phantoms in the realm of shadow caused constant provocations, violating the limits of normative experience. In 1606, the Belgian Jesuit Johann David thought to illustrate the difficulty involved in representing a normative Christ, and he exemplifies this in the persons of ten errant painters, commissioned to paint Christ bearing the cross (fig. 1). Only one of the ten painters successfully renders a faithful representation of a model standing on a hill before them. He alone recognizes the subject and renders it normatively. Most of the painters misunderstand the subject, representing other Christological

Fig. 1
Frontispiece to Johann David S.J. *Orbita Probitatis Ad Christi Imitationem Veridico Christiano Subserviens* (Antwerp 1606).

schemata instead, such as The Marriage at Canae, the Sacra Conversazione or the Resurrection. They have missed the point. In fact, two of the painters commit a grave error, painting animals' and devils' heads. They heed only their imaginations, which can beget only monsters. Imagination, the least trustworthy means of perception, perverts all normative experience and recognition (cf. Reinitzer 1983, No. 1, p. 15).

The decades around 1600, marked by the victory of the Counter-Reformation and setbacks for humanistic thought, also brought a fixed idea of the imagination. Imagination was viewed as a power of opposition, to be contained and contended with; it was identified and could be personified.

Imagination is pictured for the first time in the third edition of Cesare Ripa's *Iconologia* of 1618 (fig. 2; Ripa 1618, 603). A protégé of a Roman cardinal who had taken part in the Council of Trent, Ripa interprets his figure of imagination as follows. The multicolored drapery of Imagination corresponds to the multiplicity of external appearances that affect her. Her ragged hair and the wings at her temples indicate rapid and sudden activity, relentlessly spurring her on. This description clearly suggests a fear of the eruptive, uncontrollable and continual activity of the imagination, which still troubled sleep with its monsters and chimeras.

In 1603, Ripa's second edition still personifies imagination in a feeble and harmless form, under the heading of *capriccio*; these are ideas, Ripa says,

Fig. 2
"Imaginazione," from Cesare Ripa, *Iconologia* (Padova 1618[2]).

which manifest themselves in painting, in music or in other places out of the ordinary (*si manifestano lontane dal modo ordinario*). Imagination's inconsistency is expressed by her youth, her multiplicity by numerous colors. The feathered hair signifies that unusual actions generally originate in the imagination (*Mostra che principalmente nella fantasia sono poste queste diversita d'attioni non ordinarie*); the spurs and the bellows signify that the capriccioso is always ready to flatter virtue and censure error (Ripa 1603, 48).

For seventeenth-century artists, the *capriccio* genre encompassed a series of disconnected and chance inspirations. Such improvisations were first assembled by Callot under the name *capricci* in 1617. He presented fifty small and sketchy figures, landscapes, fold scenes, street prospects and squares, with no particular program. That these were not products of the imagination in the strictest sense of the term is indicated by the absence of monsters and chimeras; but motives and fillers were still present which usually animated gaping intervals and backgrounds of historical paintings, still favored in contemporary cut-out techniques. In these *capricci* they emerge from the old, dark backgrounds, to enjoy prominence in the light of day. Callot probably derived his concept of *capriccio* from musicians and composers like Michael Praetorius, who defined the term in 1619 with the following words:

> *Capriccio seu Phantasia subitanea*: If one attempts to treat a fugue according to his own pleasure and desire, and does not persist long in this, but rather turns soon to another fugue as it comes to mind or occurs to him ..., one can also take pride in such fantasies or capricci in his art and artificing (Praetorius 1619, III: 21; quoted in Halm 1954, cols. 330–31).

After Callot, the title *capriccio* appears repeatedly as a designation for graphic series that ascend now and again into the world of monsters and chimeras. The most famous of these are Stefano della Bella's *Diversi Capricci* of ca. 1648; the first edition of Piranesis *Invenzioni capricci di carceri* of ca. 1745; Tiepolos *Varij capriccij* of 1749 or his *scherzi di fantasia* or François de Cuvilliés *Morceaux de Caprice*, of ca. 1750 (fig. 3). These works all conform to the definition formulated by Baldinucci in 1681:

> Our artists claim to create by imagination and caprice when they produce works without a model (*senza exemplo*) and of their own invention, fully contradicting the mimesis of nature (Baldinucci 1681, quoted in Halm 1954, ibid).

Executed with reference neither to a model nor to nature, these "harmless configurations" were inherently defiant, even if they tended to remain on the

Fig. 3
Frontispiece to François de Cuvilliés, *Morceaux de Caprices à divers usages*, Book X (Munich / Paris, ca. 1750).

surface and not to delve into the realm of shadow in the name of imagination. In 1679, Imagination appears allied with the powers of darkness on the title-page of a handbook of glass-making (fig. 4). A second edition of this book, which appeared in 1743, presents a detailed explanation of the illustration: on the left, Reason and Experience illuminate the Realm of Truth, where *lux veritatis* reigns. Opposed to this on the right one sees the Realm of Darkness harboring Imagination, who blows empty breasts of *amentia* as a grotesque ornament into the lap of Understanding, inscribing them with the names of all those who err, producing phantoms. A troublesome youth pinches her foot, while the background figure of a donkey-eared Pegasus testifies to the blind faith and impudence — *temeritas* — which Imagination obeys (Mittig 1973, cols. 1453–54, fig. 2). Reason's enlightened ascent in opposition to Imagination is openly proclaimed here.

Ratio was to ban imagination in much the way courtly conventions were wont to. Painting in 1691 at the royal court of Sweden, David Klöckner Ehrenstrahl pictured himself between Pictura and Fantasia (fig. 5). The latter holds a sheet calling upon the artist to immortalize the fame of his majesty in paint. Klockner himself states that the mask in the hand of the

Fig. 4
Frontispiece to Johann Kunckel (von Löwenstern), *Ars vitraria experimentalis* (Nuremberg 1744).

Fig. 5
David Klöckner Ehrenstrahl, *Selfportrait* (1691), Stockholm, National Museum.

ascendant guardian angel signifies the unlimited capacity of painting to bring forth multifarious motifs and hence reminds one of the possibility of inventing monsters. And yet the chain held in the hand of the same angel means, as the artist says, that a good painting must maintain integrity and consonance in the composition of its parts (cf. Berghaus 1977/78, p. 9).

This maxim of propriety expressly refers to decorum, which provides the means to restrain the last evasive residues of the imagination. If Reason assumes this function, the lot of Imagination is little improved, as Carlo Orlandi demonstrates in his 1765 version of Ripa's *Iconologia* (fig. 6). A slave holds a crown over Imagination's head, no longer alluding, according to Orlandi's commentary, to phantoms but rather to the *rette idee*, i.e., to the well-reasoned and normal productions of the artist. The slave's left hand, however, carries a bunch of sphinxes and chimeras. But here, with a sweep of the hand, Imagination disclaims her traumatic progeny, turning instead toward her new protector and mentor. A lynx now attending her alludes to clarity of vision; a scepter with a seeing eye, wielded in her right hand, symbolizes the *intelletto*, clarity of understanding (cf. Mittig 1973, cols. 1455–56). Imagination has been converted to faith in Reason, and the light of reason will liberate her from her age-old trauma by rendering those chimeras superfluous.

In the same year, 1765, Diderot introduced his *Encyclopédie*, placing imagination in another constellation (fig. 7). On the title-page of 1765,

Fig. 6
"Fantasia," from Carlo Orlandi/ Cesare Ripa, *Iconologia* (Perugia 1765).

Fig. 7
Frontispiece from the *Encyclopédie*, by M. Cochin, 1765.

designed by M. Cochin, Reason and Metaphysics unveil Truth, and Imagination adorns her with garlands.[5] Although this personification still has winged temples, the chimeras have been superseded by tongues of fire. Imagination serves merely to decorate Truth. This assignment is reaffirmed at the beginning of the French Revolution in the Cochin/Gravelot compendium of iconology (ca. 1791), which characterizes Imagination as a light-hearted and elegant lady of utilitarian intentions — according to the accompanying text (fig. 8). Fantasy, or Imagination, is defined as the ability to tap the memory and bring its ideas to light. Her youth is explained by her constant occupation with *productions nouvelles*, like the figures hovering above her head. The hybrid figurations wandering through the background are explained: Figures such as centaurs, sirens and harpies owe their existence to poetry as well as to painting; they are inventions whose only justification is as emblems for truth, *vérités*, whether these be of a physical or a moral nature (Gravelot-Cochin, III, no. 1). In this reading, the old license to release monsters from their dark refuge loses its mystique and becomes a moral instrument. Such license is granted only in the service of norms.

Francisco Goya was confronted with this restriction as he designed his *Caprichos* (1799). Fluttering about the head of the dreaming author are those

Fig. 8
"Imagination," from Hubert-François Gravelot, *Iconologie* (ca. 1790).

terrifying creatures that had belonged to the cache of the imagination for centuries (fig. 9). A commentary, perhaps by the artist himself, explains:

> Imagination abandoned by Reason begets incredible monstrosities, although Imagination allied with Reason is the Mother of Art and the origin of Art's wonders.[6]

This implies that Goya might have believed a reconciliation of the two powers possible. It is significant for the history of the idea imagination that the chimeras and monsters attached to its head are now set free. Such is the situation of the modern and avant-garde artist who, relying on himself alone, would like to be free from responsibilities toward his given world.

Five years before Goya, E.T.A. Hoffmann pictured a similar situation in a gouache (fig. 10; now lost). Imagination, a stately woman with the traditional crown and winged temples, approaches the author, who has risen from his chair in order to view — not himself — but two friends in a mirror that Imagination is holding before him. This personal reminiscence takes place in a scene haunted by chimeras and fabulous creatures. The apocalyptic figure of Imagination approaches the author, offering him new and other worlds

Fig. 9
Goya, *El sueno de la razón produce monstruos*, Capricho 43, 1796.

Fig. 10
E.T.A. Hoffmann, "Imagination appears to the poet," Gouache, 1794.

(cf. Muhler 1965, 91f.; Mittig 1973, col. 1457). Like Goya, Hoffmann suggests that the world of an artist is a world apart and totally self-reliant, a world of phantoms and monsters. That other imagined world of hermetic isolation allows the artist to evade all the normative measures imposed by a bourgeois existence. Within our history of images, the modern artist liberates his subjective imagination, releasing the repressed phantasmagoria of centuries for us to contend with. Thus, shocking their public, each new wave of avant-garde artists manages to draw from that imagined fundus and produce ever new phantoms. Art has had to pay a high price for the liberty taken by its practitioners. For they threaten to take their means with them out to the margins and into realms divorced from our modern social experience. Art as a whole is maneuvering itself into a banishment long *imposed* on the imagination.

The course of this development is demonstrated by allegories of the imagination in official and academic art. They represent Imagination as a woman, radiant and untroubled by ill thought or phantoms. On the pedestal of the Beethoven monument in Bonn, erected in 1840 (fig. 11), Imagination rides a sphinx through the air; monsters and chimeras no longer flutter around her head — they have been domesticated, to serve as a means of transportation. This optimistic variation is rendered in marble in the Schubert monument on the Ringstrasse in Vienna (fig. 12; cf. Mitting 1973, cols. 1459–60; Krause 1980, fig. 41). And in the newly reopened opera house of Dresden, Phantasy was allowed to occupy her throne and accept the homage of the arts (fig. 13; Magirius 1985, pp. 146ff.).

In 1863, Hermann Wislicenus varied this ascendant Imagination showing only a few of her hybrid creatures fluttering behind her. Wislicenus entitled his version *Phantasie von Träumen getragen* (Imagination borne by Dreams),

Fig. 11
"Phantasie," on the Monument to Beethoven by Ernst Julius Hähnel, Bronze, 1840–1845, Bonn.

Fig. 12
"Phantasie," on the Monument to Schubert, by Carl Kundmann, Marble, 1867–1872, Vienna.

Fig. 13
Woodcut after Ferdinand Keller's sketch for the Hoftheater, Dresden: "Imagination with the Arts," ca. 1875.

intimating Imagination's once close association with feverish dreams (cf. Scheidig 1971, Abb. 33).[7] But such dreams trouble her no longer. Tamed, they transport her into ethereal regions. In this framework Imagination no longer obsesses us or delves beyond beautiful surfaces to drive creatures of the darkness into the light. On the contrary, she stands ready to ornament the surface of prosaic and imperfect existence. Imagination has degenerated into wishful thinking, so useful and instrumental as an *exzitative* in our social and productive systems that she deserves awards for repeatedly generating those productive solutions.

What remains of Imagination's original import is at most still apparent in the word "phantom," which escapes the grasp of given realities. Because the original goal of imagining has become the program of the avant-garde, their products are bound to share the fate of the old chimeras and monsters — an uneasy and uncomfortable marginal existence, an existence we cannot renounce.

Notes

1. On the history of the term in general cf. Bundy 1927; *Historisches Worterbuch*, IV, 218ff.; Summers 1981, 33–282.
2. Cennini 1943, cap. I, p. 20: "... che conviene avere fantasia e operazione di mano, di trovare cose non vedute, cacciandosi sotto ombra di naturali, e fermarle con la mano, dando a dimostrare quello che non è, sia ... Che'l poeta, con la scienza prima che ha, il fa degno e libero di potere comporre e legare insieme sì e no come gli piace, secondo sua volontà. Per lo simile al dipintore dato è libertà potere comporre una figura ritta, a sedere, mezzo uomo mezzo cavallo, sí come gli piace, secondo sua fantasia."
3. Cf., e.g., di Bicci 1976, 187, 413, and also Warnke 1982, 66.
4. Cf. Fehl 1961, 349–51; Schaffran 1960, 178–93.
5. Cf. Catalogue "Diderot et l'art de Boucher à David," Paris 1984/1985, No. 148.
6. Cf. Catalogue "Goya-Zeichnungen und Druckgraphik," Frankfurt/M. 1981, No. D 34, pp. 62–65.
7. As a kind of fantasma's retirement from fantasy, one may judge Honoré Daumier's series about "l'imagination," cf. Rump 1979, 226–47.

References

I. Primary Sources

Baldinucci, Filippo, 1681. *Vocabolario Toscana dell'arte de Disegno*. Florence.

Castiglione, Baldassare, 1972. *Il Libro del Cortigiano*. Ed. E. Bonara. Milano.

Cennini, Cennino, 1943. *Il Libro dell'Arte*. Florence.

di Bicci, Neri, 1976. *Le Ricordanze*. Ed. Bruno Santi. Pisa.

Dürer, Albrecht, 1969. *Schriftlicher Nachlass*. Ed. Hans Rupprich. Berlin.
Gravelot, Hubert-François and Charles Nicholas Cochin [ca. 1790] 1972. *Iconologie par figures*. Paris, Geneva.
Horace. *Ars Poetica*.
Praetorius, Michael, 1619. *Syntagma Musicum*. Wolfenbüttel.
Ripa, Cesare, [1603] 1970. *Iconologia*. Hildesheim, New York.
—, 1618. *Nova Iconologia*. Padova.

II. Secondary Sources
Berghaus, Peter, 1977/78. in Catalogue *"Porträt 1: Der Herrscher,"* Münster/W.
Bundy, Murray Wright, 1927. *The Theory of Imagination in Classical and Mediaeval Thought*. Urbana, Ill.
Fehl, Philipp, 1961. "Veronese and the Inquisition," *Gazette des Beaux-Arts* 58:349–51.
Halm, Peter, 1954. "Cappricio," in *Reallexikon zur Deutschen Kunstgeschichte,* III. Stuttgart, cols. 330–36.
Historisches Wörterbuch der Philosophie, ed. J. Ritter and K. Gründer. Basel-Stuttgart, 1976.
Krause, Walter, 1980. *Die Plastik der Wiener Ringstrasse* (= *Die Wiener Ringstrasse*, Vol. IX/3).
Magirius, Heinrich, 1985. *Gottfried Sempers zweites Dresdner Hoftheater*. Vienna.
Mittig, Hans-Ernst, 1973. "Fantasie," in *Reallexikon zur Deutschen Kunstgeschichte,* VI. Munich, cols. 1451–61.
Muhler, Robert, 1965. "E.T.A. Hoffmann und die bildkünstlerische Darstellung des Jenseits," *Jahrbuch des Kunsthistorischen Instituts Graz*, 1.
Reinitzer, Heino, 1983. *Biblia deutsch. Luthers Bibelübersetzung und ihre Tradition* (Ausstellungskatalog der Herzog August Bibliothek, No. 40), Wolfenbüttel.
Rump, Gerhard Charles, 1979. "Über Daumiers 'L'Imagination'," in: *Honoré Daumier. Der Bürgerliche Alltag*, Catalogue, Bonn.
Schaffran, Emerich, 1960. "Der Inquisitionsprozess gegen Paolo Veronese," *Archiv für Kulturgeschichte* 42:178–93.
Scheidig, Walter, 1971. *Die Geschichte der Weimarer Malerschule 1860–1900*. Weimar.
Summers, David, 1981. *Michelangelo and the Language of Art*. Princeton, N.J.
Warnke, Martin, 1982. "Praxisfelder der Kunsttheorie," *Idea* 1:54–71.

Works of the Imagination
A Comment

MOSHE BARASCH

Professor Warnke is well known for his studies of *grotteschi*, where he has demonstrated in detail the abundance of forms and motifs that a certain type of "work of the imagination" may produce. One cannot, therefore, but admire the ease with which he has taken us through the centuries in the history of this domain. Since I am essentially convinced by Warnke's conclusions, my few brief comments will be concerned mainly with contexts.

My first comment deals with something we may call "semantics." Though this is a matter of mere words, I feel something would be missed if some of the distinctions I should like to mention were not made. In the European tradition, in a continuous line leading from Greece — and perhaps earlier — to our own days, the notion and term *phantasia* has carried several connotations. It goes without saying that often these connotations cannot be neatly kept apart, and yet the differences are not negligible. The first meaning I should like to mention is perhaps the vaguest; but though it is hard to define clearly, it has tinged the notion with a definite tone or character and underlies many of its applications. What I have in mind is that *phantasia* or *imaginatio* indicated a creative faculty of man, his ability to shape or bring forth something that is new, not a replica of something already existing. In the third century A.D., Philostratus the Elder, a Greek intellectual and writer, composed *a Life of Apollonius of Tyana*, who was the founder of a religion of sorts. In the work, Apollonius expresses his views on various subjects, mainly in the form of dialogues with his devoted disciple, Damis. One of the dialogues touches inter alia on painting and on how a picture is produced.

E. Ullmann-Margalit (ed.) The Scientific Enterprise, 117–121.

Painting is of course defined as an art of imitation, but the wise Apollonius also asks about "the things which are seen in heaven whenever the clouds are torn away from one another, I mean the centaurs and the stag-antelopes, yes, and the wolves too, and the horses, what have you got to say about them? Are we not to regard them as works of imitation?" (II, 22). The origins of these images, it is concluded in the dialogue, are not in God or in Providence, nor in fact in chance, but rather in ourselves. It thus follows that it is we who possess an image-making faculty. Though Apollonius does not use the term here, this passage is actually about *phantasia*.

Just to indicate the tremendous range of this view, let me remind you of the seventeenth-century philosopher Hobbes, who, it has been claimed, was the first Englishman to take the notion of creative fancy seriously. He criticizes the contemporary views, according to which "the senses receive the species of things, and deliver them to the common sense; and the memory to the judgment, like handing of things from one to another" (Hobbes I,11). Hobbes cannot accept this picture of mental life as a toy-exchange. Fancy, he indicates, is in a sense the beginning of a chain. He still hesitates to employ the verb *creare*, to create, for an activity performed by human beings, but it is obvious that this is what he has in mind, and that the organ performing this function is Fancy.

Another connotation of *phantasia* is a heavy leaning toward visual experience, particularly the seeing of spiritual realities. Originally, of course, the term meant mainly seeing. Phantasia, we read in the Liddell-Scott *Greek-English Lexicon*, means "a making visible, a display." The Latin term is *ostentatio*. Only in philosophical language according to the *Lexicon*, does the term mean "the power of the mind by which it places objects before itself." Already Quintillian, the teacher of rhetorics, knew the dangers of imagination. That faculty, which conjures up "images by which the representation of absent objects are so distinctly presented to the mind that we seem to see them with our eyes and to have them before us," is a "lawless power." Disciplining it grants speech (and we can add, art) the power of conviction. "Shall we not turn this lawless power of our minds to our advantage?" (VI, 2, 29–30). This ability visually to perceive things absent, or things never seen, the faculty to experience optically spiritual realities, contributed a great deal to the Renaissance notion of *giudizio dell'occhio*, and through it to modern views of the artist and the poet.

Now, this combination of the sensual (vision) and the spiritual is not restricted — nor, in fact, necessarily linked — to the deformed and the contorted. The idea of the object seen in fantasy, the idea of a nonexistent object internally perceived, hovering between matter and mind and linking them, played a crucial part in the intellectual history of Europe. It could also refer to a highly idealized reality. Gerhart Ladner has called our attention to an interesting parallelism. Explanations of *imaginatio* in the twelfth and early thirteenth centuries make it the meeting point between the basic realms of being, matter and spirit. "The spirit and the flesh can be easily united," he writes on the basis of scholastic views, in particular those of Issac of Stella, "where their extreme borders touch, that is to say, in the imaginative realm of the soul..." (*Ad Imaginem Dei*, pp. 55ff.). The similarity to the idea of Incarnation is unmistakable. In fact, says Ladner, the twelfth century brought both an intensive concern with *imaginatio* on the part of schoolmen, and an intensive concern with the image of Christ on the part of artists inspired — or guided — by schoolmen. Whether or not you accept this interesting suggestion, it clearly shows that fantasy was perceived as a highly idealizing, one might say — redeeming, function.

The last aspect of *phantasia* I wish to mention is the one Warnke has discussed in his paper: the embodiment of imagination in what — for want of a better term — we often call "the fantastic." I mean the hybrid, the contorted, the unlikely. I do not think I can add anything new except for stressing once more that this is only a segment of how fantasy was actually understood. But, as we have learned, it is a highly significant segment, placing the creative character of imagination as such before our eyes.

These semantic observations lead me almost directly to my second comment, which is no longer concerned with semantics, that is, the usage of terms and distinctions between different shades that go under the same name: we are now confronted with a question of meaning. I have often wondered — and Warnke's paper has further intensified my amazement — about a seemingly paradoxical situation. Like Warnke, I am convinced that "the fantastic," the deformed and the contorted, the unheard-of and the unbelievable are indeed typical embodiments of the very essence of the imagination in general. But if this is so, one cannot help asking why those works of the imagination, those fantastic creatures banned to the margins of manuscripts, are placed — often actually hidden — in the less prominent areas of churches, where visibility is poor; why they are so often placed so high up on columns or walls that it takes the art historian's curiosity (and a very good camera) even to notice them. Moreover, in later periods, fantastic creatures tended to populate the frames rather than the centers of pictures. Even nowadays, the works of surrealism, the many forms of fantastic distortion so often found in the visual production of our age and world, elicit a response somewhat different than the works of most other styles. Why is this so?

We are familiar, of course, with the answers usually given to similar questions. In many periods, it is claimed, the artist's social position was not high enough to justify the display of his inner experiences; for other periods there are different answers, which I cannot discuss here. These answers are surely correct, as far as they go, but for our particular purpose I am afraid they do not go far enough. Here I shall make, though with some hesitation, the rather bold claim that in European culture (and perhaps this is a general human trait) the attitude toward imagination is ambivalent. "Works of the imagination" do indeed stand for imagination in general. Therefore, in rejecting and suppressing these wild "works of the imagination," periods and cultures were actually trying to suppress fantasy as such, to deport it into invisible corners, ridicule it or transform it into mere "playfulness."

It is fascinating to follow the history of the profound suspicion that Western culture persistently entertained toward imagination. I have already quoted Quintillian's opinion that *imaginatio* is a "lawless power." He also describes the imagination as a *vitium animi*. I cannot help referring to St. Augustine, who not only exerted a profoundly formative influence on the intellectual and emotional climate of many centuries, but was also sensitive to the arts and their expressive power as only few thinkers have been. St. Augustine sensed that in the *imaginatio* a radical freedom of man is implied. Because he so forcefully sensed the freedom of imagination, he was so totally afraid of where that freedom might lead. "Thus," he writes in one of his epistles (*Ep.*7), "according to my own fantasy, and as it may occur to my own mind, I picture to myself the appearance of Aeneas, or of Medea with her team of winged dragons, or of Chremes, or Parmeno." The emphasis here is on the words "my own." One can well understand how, in the Middle Ages, the faculty of imagination came to be associated with the sin of Pride. Imagination, to quote David Summers, a contemporary art historian, is a "Promethean virtue." The association of painting, poetry, license — or imagination, for which license is needed — and pride is an ancient one. It also had a long and significant ancestry in the Florentine tradition. In *Purgatorio*, Dante describes Pride, like Imagination, as two-faced: On the one hand, it is a spur to high human accomplishment; on the other, it is a source of mortal peril to the soul. It is to this very same tradition that Cennini belongs.

When the painter's fantasy was attacked by the representatives of religion, this was not always done on behalf of some kind of "normality"; often, or mainly, it was also intended to ensure that man, for whom the artist stands here, would not rebel against God. Let me add some details from the Florentine tradition. In 1496 Savonarola, preaching on the art of dying, said, following the scholastics, that to understand something one must form a "phantasm" of it in one's mind. These phantasms, however, are "eyeglasses of the intellect." They can tinge what one sees with a certain tone, but they cannot produce what one sees. The creation of a reality out of nothing, we understand, remains God's work alone.

The well-known humanist Juan Luis Vives praised *fantasia* precisely for the unrestrained freedom that is its essence. *Fantasia*, he said, is "prodigiously unrestrained and free; it can form, reform, combine, link together and separate; it can blend together the most distant objects or keep apart the most intimately connected objects." But this radical freedom, as we know, is also a mortal peril for man. Let me conclude with the well-known lines from a poem Michelangelo wrote in his old age, when he repudiated the spiritual substance of his art:

> Onde l'affettuosa fantasia
> che l'arte mi fece idol e monarca
> conosco or ben com'era d'error carca
> e quel c'a mal suo grado ogn'uom desia
>
> (Whence the impassioned fantasy
> that made art my idol and monarch
> I now know well how full that was of error
> and that which each desires to his loss)

I need not go on adding examples. The twilight in which fantasy was seen, the ambiguous attitude toward it, are obvious. The association of sinfulness was one the notion of fantasy could not rid itself of until very recently. And it may well be argued that, in some suppressed and transformed way, it survives even in our own generation. Can it not be, one asks oneself, that the deformed and contorted shapes traditionally associated with the "fantastic" reflect this survival? I think it is not altogether a matter of chance that the figures regarded as the embodiment of creative fantasy so often have an affinity of character with the features traditionally considered characteristic of the devil and of those possessed by him. What is expressed in these contorted figures is not only human creativity and playfulness, but also the bipolarity of imagination, still alive today.

Wittgenstein, Following a Rule, and Scientific Psychology

J. J. C. SMART

This paper's topic is Wittgenstein's remarks about following a rule and whether they are compatible with a scientific psychology. There will in fact be little about scientific psychology as such in it, but I hope that my remarks will bear upon this question, even though indirectly. At least, this is my motive in discussing what Wittgenstein and some of those who have commented on him have said. The subject is one on which I had no settled views, and I thought that it would be a good thing to try to make up my mind on the matter. At any rate, the prospect of presenting the paper has been a powerful stimulus to reading and thinking about the subject. Whether I shall have anything new to say about it is another matter. There have been differing views about what Wittgenstein had to say and about what he ought to have said. I have not managed to read all the literature on the subject, but from the considerable amount that I have read I get the impression that practically everything that could sensibly be said on this topic, whether true or false, must have been said by someone, somewhere or other. Having seen in the *Times Literary Supplement* (24 March 1985) Avishai Margalit's very good review of Colin McGinn, *Wittgenstein on Meaning* (1984),[1] and of G. P. Baker and P. M. S. Hacker, *Scepticism, Rules and Language* (1984), I wondered whether in coming to Jerusalem I would not be bringing "coals to Newcastle."

One striking thing about the vast literature about Wittgenstein on following a rule is the considerable disagreement about what Wittgenstein's position really was. For example, Kripke (1982) has one interpretation, and McGinn

E. Ullmann-Margalit (ed.) The Scientific Enterprise, 123–137.

has quite a different one. I am inclined to side with McGinn here against Kripke. Even so, Kripke's Wittgenstein might be just as interesting as the real Wittgenstein. McGinn therefore quite rightly has a chapter discussing Kripke's Wittgenstein in its own right.

I sometimes wonder whether the wide variety of interpretations of Wittgenstein on this topic may come from an attempt to see something much more profound than is actually there. We must remember that when Wittgenstein was writing what became the *Philosophical Investigations*, the *Remarks on the Foundations of Mathematics*, the *Zettel*,[2] and so on, it was very much in the philosophical air that the topic of the psychological was the topic of *consciousness*, so that understanding or meaning something, for example, was supposed to be an episode in consciousness. (Even the Freudian subconscious was thought of as resembling the conscious, but inaccessible to our awareness, rather like the dark depths of the ocean.) It was this sort of view that Wittgenstein was concerned to refute. Philosophers happily talked about acts of belief, and of motives, desires, and emotions as though they were occurrent feelings, of meaning as imagery, and so on. Wittgenstein, and also Ryle, rid philosophy of this sort of confusion so successfully that much of what they said so excitingly at the time has come to seem pretty platitudinous to the younger generation of today.[3] This, of course, is to some extent as it should be, if philosophy is to emulate the sciences in being a cumulative enterprise. Wittgenstein is concerned to deny that following a rule, meaning something by our words, or understanding, is an episode in consciousness, or something essentially brought about by an episode in consciousness. In part, he is saying something that was said with less fuss by Gilbert Ryle, who treated desires, beliefs, and so on as dispositions to behavior, not as events or processes in a Cartesian consciousness (Ryle 1949). (Of course, one must ask how much Ryle himself was influenced by Wittgenstein.)

Thus, consider Wittgenstein's example of the person who enumerates the sequence 2, 4, 6, ..., 1000. The natural way for this person to go on would be: 1002, 1004, 1006,... One could of course go on to 1004, 1008, 1012,... What is it to be going according to one rule rather than another? Wittgenstein wants to reject various answers. If these were the only answers one would get into skeptical paradox (PI, 185ff.). According to Kripke, Wittgenstein was actually arguing for a skeptical paradox, to which he gave a "skeptical solution," analogous to Hume's skeptical solution to the problem of induction; but I agree with McGinn and with Baker and Hacker that this was not Wittgenstein's intention.

The sort of answer that Wittgenstein wishes to reject is that there is some episode in consciousness of *meaning* a particular rule. For it might be said that it consists of having a mental image of an inscription of a formula or verbal instruction. Wittgenstein replies that such an image in consciousness itself needs interpretation, and this means following a rule: the rule for interpretation of the image. And, on the traditional view that he is concerned to attack, this clearly leads us on to an infinite regress. Wittgenstein agrees, of course, that is is more natural to interpret the formula "$a_n = 2n$" as instructing us to continue 1002, 1004, 1006, ..., and not 1004, 1008, 1012, ..., but he makes the point that there is no uniqueness of possible interpretation. Consider a drawn arrow, which we might take either in the normal way or as an instruction to move in the direction from head to tail (cf. PI, 454). If an episode in consciousness were interpreted by means of another episode in consciousness, this second episode would itself need interpretation, and so on with the regress. Wittgenstein takes this as a *reductio ad absurdum*, to show that meaning something (e.g., one continuation of a sequence rather than another) cannot be an episode in consciousness.

There are of course episodes in consciousness that occur when we are coming to understand something (e.g., a rule). But coming to understand is not understanding. Ryle held that the former is a process, while the latter is a disposition. (Wittgenstein did not like talking of "dispositions"; I shall come to this shortly.) Ryle's distinction between "task" or "getting it" words and "achievement" or "got it" words can be used to reinforce Wittgenstein's argument. As Ryle pointed out, this distinction was anticipated by Aristotle (*Metaphysics*, 1048b 30–34). There are all sorts of events and processes that can occur when one is coming to understand, just as there are in coming to discover (Wittgenstein would perhaps say "invent") a mathematical proof. Many of these will doubtless be going on in the brain, of which we are unconscious, but we are commonly aware of many of these episodes. Because of his stress on consciousness and playing down of *brain*, when he talked about the mind, Wittgenstein (in PI, 151) mentions only those of which we are aware. He gives the example of a man *A* who has written down the numbers 1, 5, 11, 19, 29. Another person *B* comes to understand how to go on. What happened in *B*'s mind? As Wittgenstein points out, various things might have happened. For example, *B* might try out various formulas, and as *A* got to 29 he or she tried the formula $a_n = n^2 + n - 1$, which he or she would find confirmed when *A* proceeded to write down 41 (PI, 151). There is no necessary condition of this sort for coming to understand. Compare journeying and arriving. A person traveling from London to Edinburgh might go via Berwick on Tweed, or by Carter Bar, or by Carlisle, as well as by a variety of other routes. None of these is necessary. Arriving is the success or result of journeying. Understanding is the success of coming to understand, though the parallel is closer if we compare understanding not with arriving as such but with the state of being at one's destination.

In PI, on p. 154, Wittgenstein says "Try not to think of understanding as a 'mental process' at all. For *that* is the expression which confuses you." It seems to me that Wittgenstein could have said here, as Ryle would have, that it is not so much that understanding is not a mental process, but that it is not a process at all. As Ryle would insist, one could not say that one was halfway through understanding the French for "cow" or that one was hard at work understanding the French for "cow." A student might perhaps say that he or she was halfway through and hard at work at understanding a difficult proof in logic, but the student would have spoken more accurately if he or she had spoken of being halfway through or hard at work at getting to understand the proof. Why did Wittgenstein not allow that understanding was a mental state or disposition? Three reasons can be suggested.

First, Wittgenstein refused to call mental states "dispositions" because these would primarily be dispositions to behavior, and there is no necessary and sufficient set of behavioral conditions — no translational analysis of belief, desire, understanding, and so on into hypotheticals (including counterfactuals) about behavior. (Here there is merely a verbal difference from Ryle, whose use of the term "disposition" was to indicate, in a way that fell short of precise translation, sets of behavioral hypotheticals.) An answer to Wittgenstein's objection could be that mental dispositions are structural states of a person (perhaps contingently identified with brain states) that *explain* the behavioral hypotheticals. In fact, they would do so not on their own but in conjunction with other such states. To believe that it will rain is not a simple disposition to take a raincoat, because one may not mind getting wet, but the belief, together with the appropriate desire, explains the behavior.

The second reason why Wittgenstein refused to identify understanding with a disposition or a state of a mental apparatus explanatory of behavior is that, if it were, "there ought to be two different criteria for such a state: a knowledge of the construction of the apparatus, quite apart from what it does" (PI, 149). Why can one not postulate the existence of a brain state as explanatory of behavior even though one does not know (mainly because of the complexity of the nervous system) how to describe the neurophysiological details of this brain state?

When Wittgenstein discounts thinking of following a rule as behaving in accordance with some causal mechanism, he discounts the possibility of a ghostly mechanism. But why not a physiological mechanism? In discounting this possibility Wittgenstein is perhaps motivated by his desire to separate conceptual questions from empirical ones. Of course, those of us who have been convinced by Quine's objections to the analytic-synthetic distinction would not wish to separate these types of question so sharply. However, I shall not press this point here. In Z, on p. 608, Wittgenstein even plays with the idea that *nothing* may correspond physiologically to a mental process. He says, surprisingly, that "No supposition seems to be more natural than that there is no process in the brain correlated with associating or thinking; so that it would be impossible to read off thought-processes from brain-processes." This might perhaps be taken merely as suggesting a token-token identity theory rather than a type-type one, but this does not seem to be in Wittgenstein's mind, because he goes on to suggest that our intelligent behavior might arise out of chaos in the central nervous system.

Suppose that we found that an apparently normal person's head was full of sawdust where we expected to find gray matter. McGinn says that in entertaining this supposition we are not committed to a real metaphysical possibility (1984, 114). Here his notion of metaphysical possibility seems to be that of Kripke's *Naming and Necessity* (1980), to which he refers. I do not myself want to go along with Kripke here, and I would prefer in this connection to talk of scientific possibility. (Kripke's conception seems to me to depend on dubious modal intuitions.) Since I believe metaphysics to be continuous with science, so that metaphysics is the more conjectural and conceptually puzzling end of science,[4] the difference may not matter much.

What Wittgenstein *might* be saying in Z, on pp. 608–10, may merely be that is is *logically* possible for a person to have thoughts and yet have a skull full of sawdust. Though a twentieth-century brain surgeon would not believe his or her own eyes if he or she opened a patient's skull and found only sawdust, would go to a psychiatrist, or resign from the hospital on grounds of insanity, Aristotle would not have been so concerned. Believing the brain to be an organ for cooling the blood, he might have thought sawdust not too bad for doing the job. Possibility is relative to contextually agreed background knowledge. (*Roughly*: "It could be the case that *p*" is truly asserted if the proposition that *p* is consistent with contextually agreed background assumptions.) In reading fairy stories, for example, we must put much knowledge out of our minds, and Aristotle had less physiological background knowledge than we moderns have. Logical possibility is the limit in which we put all background knowledge to one side, save for knowledge of logical consistency.

Because of the context dependence of our use of words like "can" and "possible" it is not clear whether Wittgenstein in Z, on pp. 608–10, is talking of mere logical possibility, or whether he thinks that what he says is consistent with accepted scientific fact. It is hard to say. There are antiscientific overtones to some of what he says, but perhaps we should charitably interpret him as speaking of logical possibility only. The ambiguity recurs in Z, on p. 611: "The prejudice in favour of psychophysical parallelism is a fruit of primitive interpretations of our concepts. For if one allows a causality between psychological phenomena which is not mediated physiologically, one thinks one is professing belief in a gaseous mental entity." Is he saying that there is no such causality — no causal explanation of how the brain affects behavior, or is he merely rejecting dualism?

A third reason which may have been behind Wittgenstein's reluctance to allow that understanding could be a state or disposition (which could plausibly be identified with a brain state) is connected with his idea that philosophy is a purely conceptual investigation, quite independent of empirical facts. Wittgenstein wants to separate philosophy from science. (This goes back to his *Tractatus* period: "Darwin's theory has no more to do with philosophy than any other hypothesis in natural science"; *Tractatus* 4, 1122.)

Ryle had a similar concern, though in his case it was also connected with his concern for philosophy's niche as a *professional* specialty. (Wittgenstein did not really approve of philosophy as a professional activity.) Ryle was also concerned to debunk the psychologistic view of philosophy (epitomized by the title of the journal *Mind*) and also the Platonizing view of philosophy. He had some regard for the latter conception in that it opposed psychologism, but he deplored its positive, other-worldly aspect. Ryle found a niche for philosophy as "talk about talk," though not as a philological enquiry but as a conceptual one. In Ryle's case, perhaps more overtly than in Wittgenstein's, there was an antiscientific undercurrent. This was partly just modesty: he felt unqualified to talk about science because of his lack of knowledge of the specialties concerned. Nevertheless, some of the things he said did suggest a skepticism about a scientific psychology, and even led M. H. Pirenne plausibly to interpret him as a biological vitalist (Pirenne 1951–52). I suggest that in Wittgenstein also there may be the same combination of official scientific neutrality and an underlying hostility.

An openness to vitalistic ideas seems to be implicit in Z, on p. 613: "Why should there not be a natural law connecting a starting and finishing state of a system, but not covering the intermediary state? (*Only one must not talk of causal efficacy.*)" I find the last sentence of this quotation obscure, but the earlier part certainly suggests the (not merely logical?) possibility of laws relating to the behavior of organisms not explicable in terms of neurology, biochemistry, and so on.

It is, of course, fair to say that these vitalistic and antiscientific tendencies in Wittgenstein are not essential to what I conceive to be Wittgenstein's main theme. This I take to be his rejection of an account of "understanding," "meaning," and "knowing how to go on" in terms of events and processes in consciousness, or of an inner ghostly mechanism. Understanding a rule, according to Wittgenstein, is not to have something occur in consciousness, such as some image or feeling, but to have a set of abilities . One can agree with all this and yet take the view that such abilities should in the end be explicable in terms of some physiological mechanism.

Consider a computer programmed to print out the sequence <2, 4, 6, ..., 1000, 1002, ...> (up to some undetermined but fairly large number; obviously, no real computer could go on forever). We can account for this printing behavior of the computer mechanistically from knowledge of the physical embodiment of the program and of the circuitry of the computer hardware. Similarly, if the computer prints out the sequence < 2, 4, ... 1000, 1004, 1008, ... >, this is explicable by reference to the physical embodiment of a slightly more complex program. Does it seem too farfetched to say that in the former case the computer "knows how" to go on one way, and in the latter case it "knows how" to go on another way? It would be too much to say that such a simple computer "understands" how to go on. Understanding implies a multiplicity of abilities. It presumably requires the abilities to use words such as "number" and "sequence" in various contexts. It is by no means obvious that there is any reason in principle why, with the development of artificial intelligence, an electronic artefact should not manifest such a cluster of abilities.

It may be objected that the computer is not following a rule, because *whatever* it does will be to enumerate some sequence or other. We do not have any notion of *correctness* in the computer's behavior, except parasitically on the intentions of a computer programmer. This, it will be said, introduces a normative element into the notion of rule following.

Certainly in some contexts the notion of rule following is social and normative. Indeed, this is a consideration that lies behind Kripke's interpretation of Wittgenstein. Sometimes the notion of a rule comes in a social context, in which we tend to talk not so much of someone following a rule but of acting in accordance with a rule. Here the notion of correctness or incorrectness is indeed a normative one. But this is not the *explanatory* context, in which we say that a person wrote down 41 because he or she was following the rule $a_n = n^2 + n - 1$.

I think we can elucidate a nonnormative notion of correctness that could be applied to a computer. Suppose that a computer went 2, 4, 6, ... , 1000, 1004, 1006, ... Could we say that it had behaved incorrectly, missing out the term 1002? Could we not equally as well (or as badly) say that it was obeying the rule $a_n = 2_n$ if $n < 500$ and $a_n = 2n + 2$ if $n > 500$? I think it depends on the circumstances.

Suppose that the program of the computer is such that the computer will enumerate 2, 4, 6, ... , 1000, 1002, 1004, ... , but it misses 1002 because of a burst of cosmic rays, or a lightning strike, or some other random outside interference. Correct behavior, in this case, is what it will do if it remains an effectively closed system, so that there are no outside influences that either reprogram or alter its hardware. Similarly, we can regard behavior due to wearing out of components and exhaustion of memory store as incorrect.[5]

Of course, a machine may enumerate correctly even when subjected to random outside influences. Consider a radiation counter which behaves correctly if and only if its disposition to enumerate a sequence of numbers is affected by random outside influences. The notion of correctness here would be relative to the intention of the designer or programmer. I am not sure how to eliminate this sort of case, but I am inclined to think that something of a nonnormative notion of correctness might be developed on the lines I have suggested. Thus, the radiation counter is not following a rule, since the numbers it writes down are due to random influences.

As I have said, there is certainly a normative use of the expression "following a rule." But in the context "Why is this person going 2, 4, 6, ... , 1000, 1002, 1004, ... ?" the answer "Because this person is following the rule 'Start with 2 and keep adding 2'" is an explanatory and not a normative remark. Moreover, the explanation is equally good whether or not the rule is in accordance with the norms of a social group. A similar explanatory, nonjustificatory, use of the concept of correctly following a rule can apply to a machine, where the notion of correctness is concerned with absence of random external or internal influences. Thus, the wearing out of a component would be a random influence of the sort that might lead us to speak of the machine's behavior as being "incorrect" in such a nonnormative sense.

Earlier in this paper I indicated some reasons why Wittgenstein refuses to ground hypotheticals about behavior in categoricals about persons (cf. McGinn 1984, 34). Thus in BB, on p. 113 and p. 115, he says that we should think of abilities in terms of the tests for them. It seems to me that the only grounding for such a behaviorist concept of ability would be a Humean inductivist projection of the notion of the tests for them (as if "This *A* is a *B*," "That *A* is a *B*," "That other *A* is a *B*," ... should be thought of as *tests* for "All *A*'s are *B*'s"). This behaviorism leads him to say that it is impossible that there could be only one occasion on which a rule was obeyed. On the other hand, if following a rule was taken as a manifestation of an inner (physiological, cf.

McGinn 1984, 34) state, there could easily be only one such occasion. Consider a machine programmed to enumerate the sequence <2, 4, 6, ... >. It goes "2" and is then immediately destroyed with a hammer. Consider another machine programmed to enumerate <2, 3, 5, 7, ... >. It goes "2" and then is also immediately destroyed. There is a difference between the two machines, despite their exactly similar behavior, which would surely correspond to obeying the rule just once. Moreover, this difference is not parasitic on the behavior of other machines enumerating longer sequences.

Suppose that a machine is programmed to enumerate <2, 4, ... , 1000, 1002>, and when it gets to 1000 the program is modified so that it continues <1004, 1008, 1012, ... >. This would be a crude analogy of a person at first intending to follow the former rule but then intending to follow the second rule. It would not correspond to intending all the time to follow the rule <2, 4, 6, ... , 1000, 1004, 1008, ... >, though the total behavior would be exactly the same.

There is a fact of the matter as to which sequence the computer is enumerating, even if it is destroyed after only one move. I think that Wittgenstein would say that there is a fact of the matter as to which rule a person was following, even if this person were to be shot after only one move. For Wittgenstein this fact of the matter would be expressed by counterfactuals about what the behavior of the person would have been if the person had not been shot. As I have urged earlier, there is a problem about the backing for such counterfactuals. I would back them up by assumptions about an internal (physiological or functional) state.

Though Wittgenstein can allow that a person might obey a rule though destroyed after only one move, he denies that there should be only one occasion on which anyone ever obeyed a rule. To obey a rule, he says, is a custom, a use, an institution (PI, 199). Now the expression "to obey a rule" might be taken to refer to an *act* of obeying a rule, or it might be taken to refer to the *concept* of obeying a rule. I think that if anyone, ever, obeyed a rule only once, this would be miraculous. (The person could not have been taught the rule verbally, for example, because verbal communication itself involves extended rule following on the part of the communicators.) Wittgenstein, however, seems to go beyond thinking that a single obeying of a rule (by anyone, anywhere, at any time) would be miraculous. I think that he thought that it would be senseless in some semantic way, or as he would rather have said, it would have been a grammatical mistake. But he *may* have been saying merely that if there were only one act of obeying a rule, none of us would have the *concept* of obeying a rule. I can agree with this last idea.

I have used the analogy of programming a machine. How are we ourselves programmed? We are sometimes programmed by being given a segment of a sequence, <2, 4, 6, 8>, say, which we then extend in a way that comes naturally to us, <10, 12, 14, ... >, for example. This naturalness presumably comes from a combination of earlier influences (which we can think of as programming) and of genetically determined brain structure. Sometimes we are given a formula (such as $a_n = 2n$). We apply this formula in enumerating a sequence of numbers. We do not need to *interpret* the formula (translate it into another idiom). We might do so, but on pain of infinite regress, we must be able to apply some formula without interpreting it. As Wittgenstein says, "to use a word without justification does not mean to use it without right" (PI, 289).

Wittgenstein is correct, then, in arguing that following a rule is not having certain special events in consciousness, mental images, acts of meaning, acts of interpretation, or what not. He holds that what rule a person is following (or indeed whether he or she is following a rule at all) is a matter of what behavioral counterfactuals about the person are true. As far as I know (and I am not familiar with all of the huge posthumous corpus of Wittgenstein's work), Wittgenstein did not show much interest in the semantics of counterfactuals or strong conditionals in general. As I have suggested earlier, the counterfactuals must be grounded in postulated structural states of the person in question, and I would then contingently identify these states with brain states.[6] I have tried to indicate why this sort of move was unattractive to Wittgenstein. Nevertheless, it does seem perfectly compatible with his actual arguments, and it is indeed a natural extension of them. Moreover, it seems to me that if we do not go in some such direction, our counterfactuals are up in the air.

Up in the air or not, Wittgenstein does not seem to be skeptical about our ability to assert them, mysterious though this ability for aerial flight should have appeared to him. That is, I am following McGinn, and also Baker and Hacker, in denying that Wittgenstein was proposing such a paradox. Consider a person who goes "2, 4, 6, ... , 1000" and then stops (one has to stop somewhere). Wittgenstein holds that there is no psychological fact (by which he meant no event in consciousness) that determines whether the person has been following the rule "$a_n = 2n$" or the rule "$a_n = 2n$ if $n \leq 500$ or "$a_n = 1000 + 4(n - 500)$" if $n > 500$. Even if a person had a mental image of an inscription of one or other of these formulas, this would not settle the matter.

However, if Wittgenstein really were a skeptic about meaning, he would be in danger of ending up like Protagoras in Plato's *Theaetetus*, unable to say anything at all. Kripke's interpretation of Wittgenstein is that Wittgenstein evaded skepticism about whether there was a fact of the matter about rule following, by saying that though there was no individual fact of the matter, there is a propriety in talking of rules, where we are expressing conformity with social norms. McGinn has convinced me that this skeptical and "social" interpretation of Wittgenstein will not do. It is better, I think, to take Wittgenstein as arguing by *reductio ad absurdum* that an account of understanding in terms of events in consciousness will not do. Wittgenstein concludes that a behaviorist or "use" account of meaning, understanding, and rule following is needed.

We can get between the horns of an unacceptable dilemma. We need not give an account of rule following as conformity with social norms. (This is not to say that in nonexplanatory contexts this may be a pertinent account, and it could in certain cases be explanatory in terms of a higher-order rule of following social norms. But this of course would lead us back to the original problem.) We can also reject the other, behaviorist horn of the dilemma. Despite Kripke's objections to it (Kripke 1982, 32–37; I have tried to answer these objections by implication rather than directly) I think that the computer analogy shows that whether something is a case of rule following is a matter of dispositional state (a physiological state or, equivalently for our purposes, a functional state, since a functional state must be embodied in a physiological one). Wittgenstein had to stress the habitual character of rule following because he had to explicate it in terms of regularities of behavior, but the dispositional account in terms of states of the person (perhaps physiological ones) enables us to dispense with this reliance on habit, custom, or other such regularity.

Let me further illustrate the point by reference to the example of "grue" and "bleen," due to Nelson Goodman (1973), discussed in the present connection by Kripke and developed further in Margalit's review, already mentioned. Let 'grue' be defined as "green until A.D. 2500 and blue thereafter," and let 'bleen' be defined as "blue until A.D. 2500 and green thereafter." Goodman is concerned with a problem about inductive inference. If all our evidence consisted merely in the observation of emeralds it would seem to support "all emeralds (anywhere and anywhen) are grue" just as well as it supports "all emeralds are green." Kripke's puzzle is not about induction but about meaning. Since none of us will be alive in A.D. 2500, we will apply the word 'green' to just the same objects as those to which we would apply the word 'grue.' So the question here is whether there is a fact of the matter as to whether a person was using 'green' as we do, or whether that person is *using* it as we use 'grue.' Indeed there is a question, whether there is a fact of the matter as to whether we are 'green' speakers or 'grue' speakers. A natural objection to Goodman's paradox about induction is that 'grue' but not 'green' requires a date for its definition. (In fact, we do learn 'grue' in this

more rigid than any other" (RFM, I-119). Surely we must accept this. The theory of Turing machines is isomorphic with arithmetic — with the device of Gödel numbering a Turing machine can be defined purely arithmetically, and so can its computations. Whether a certain brain will produce a certain sequence is a fact about its design (and depending on certain assumptions, such as that subsequent events will not cause unforeseen changes in its structure). Such a fact could in principle be exhibited via Gödel numbering as a fact of *applied* mathematics.

If I have interpreted Wittgenstein's PI correctly, it still seems to me to be an open question whether PI and RFM can be completely reconciled with one another. Leaving this question to one side, and concentrating mainly on PI, it seems to me what is good in Wittgenstein on following a rule *is* compatible with a scientific psychology and indeed can be interpreted with it. We just need to excise the inessential "antiscientific" remarks and modify the behaviorism.

Notes

1. I am also sympathetic to Blackburn 1984.
2. For the abbreviations I shall use, see References below.
3. Compare remarks by McGinn 1984, 32.
4. According to this conception, of course, many theoretical physicists are absolutely first-rate metaphysicians.
5. See Goldfarb 1985, especially p. 477, and the Critical Notice of Kripke (Horwich 1984), to which Goldfarb refers.
6. Or, in this context equivalently, functional states realized by the physiological states.
7. William Grey (formerly Godfrey-Smith) has suggested a behavioral asymmetry between "green-blue" speakers and "grue-bleen" speakers. If the Normans were asked what color lettuces would be in A.D. 2500, they would say "green," whereas the Saxons would say "blue." However, this would be canceled out if the Saxons projected regularities in the "grue-bleen" way. At least, if they could similarly modify the way they talked about light rays, the chemical constitution of chlorophyll, etc. It is hard to be sure that there could be a complete restoration of symmetry. At any rate, Saxon science would have to be very complicated compared with Norman science.

References

I. Works by Wittgenstein

BB = *The Blue and Brown Books*. Oxford: Basil Blackwell. 1964.

PI = *Philosophical Investigations*. Oxford: Basil Blackwell. 1963.

RFM = *Remarks on the Foundations of Mathematics*. Oxford: Basil Blackwell. 1956.

Tractatus = *Tractatus Logico-Philosophicus*, trans. D. F. Pears and B. F. McGuiness. London: Routledge and Kegan Paul. 1961.

Z = *Zettel*. Oxford: Basil Blackwell. 1967.

II. Other Works Cited

Baker, G. P., and P. M. S. Hacker, 1984. *Scepticism, Rules and Language*. Oxford: Basil Blackwell.

Blackburn, S., 1984. "The Individual Strikes Back," *Synthese* 58:281–302.

Goldfarb, W., 1985. "Kripke on Wittgenstein on Rules," *Journal of Philosophy* 82:471–88.

Goodman, N., 1973. *Fact, Fiction and Forecast*, 3d ed. Indianapolis: Bobbs-Merill.

Horwich, P., 1984. [Critical Notice of Kripke 1982], *Philosophy of Science* 51:163–71.

Kripke, S., 1980. *Meaning and Necessity*. Oxford: Basil Blackwell.

—, 1982. *Wittgenstein on Rules and Private Language*. Oxford: Basil Blackwell.

Lewis, D., 1972. "Psychophysical and Theoretical Identifications," *Australasian Journal of Philosophy* 50:249–58.

McGinn, C., 1984. *Wittgenstein on Meaning*. Oxford: Basil Blackwell.

Pirenne, M. H., 1951–52. "Mind-Like Behaviour in Artefacts and the Concept of Mind," *British Journal for the Philosophy of Science* 2:315–17.

Ryle, G., 1949. *The Concept of Mind*. London: Hutchinson.

How to Outsmart the Rules
A Comment

AVISHAI MARGALIT

Professor Smart is a gold miner, not a coal miner. He is thus in no danger of bringing coal anywhere. As for his gold, we had better recall that all that glitters is not gold. So let me try his metal.

In his influential book on following rules according to Wittgenstein, Kripke interprets Wittgenstein as one who sees the concept of following a rule as harboring a deep paradox. The paradox is that, given two rules, R_1 and R_2, which up till now have coincided in their application, there is nothing to determine whether we are being guided by R_1 or by R_2, even though these rules are known to diverge in their future application. For example, there is nothing to determine whether I use 'green' in the sense of green or of grue (where 'grue' is what was observed before the year A.D. 2500 and found to be green, or blue otherwise). In other words, according to Kripke's interpretation there are no truth conditions to my claim that I myself am using 'green' in the sense of green and not grue. All of us, of course, feel certain that we use the adjective in one sense and not in the other, but this feeling cannot be justified on the basis of any fact whatsoever.

Smart does not share the view that Wittgenstein tried to present a skeptical paradox of the sort under consideration. Wittgenstein, according to Smart, was aiming at something more modest and more true; namely, at undermining the view that following a rule, like understanding and like meaning (something), is a mental episode. This, one should add, was a common view in Wittgenstein's time, explicitly expressed, e.g., by William James.

Two contributions are offered in Smart's presentation. (a) He offers his

E. Ullmann-Margalit (ed.) The Scientific Enterprise, 139–144.

own *interpretation* of Wittgenstein's problem about rules. (b) He offers his own *solution* to the problem of providing truth conditions for a proposition ascribing to someone the use of a rule. Smart believes that his solution contradicts no position, explicit or implicit, of Wittgenstein's. Moreover, he believes that his solution is very much in Wittgenstein's spirit, even if not in his style.

There is no doubt, I think, that Smart is right in holding that Wittgenstein was attacking the notion that mental episodes of the sort, e.g., of mental images (ideas), play a role in the explanation of the concepts of rule and meaning. For my part, I believe that Smart has located part of Wittgenstein's problem, but not its entire extension. Wittgenstein attacks not only the explanation of rules and meaning in terms of mental episodes, but also their explanation in other terms, among them dispositional terms of the very sort offered here by Smart.

However, before turning to the interpretation of Wittgenstein, let us dwell first on Smart's solution to Kripke's problem, and only then ask whether it is compatible with what Wittgenstein has to say on the matter.

There are facts, claims Smart, in virtue of which the sentence "I use 'green' in the sense of green and not of grue" is true. Smart considers two types of such facts: (a) facts regarding the disposition to use 'green', and in particular regarding the categorical (structural) base of that disposition; (b) facts relating to the way the term 'green' is learned, which in his view is very different from the way the term 'grue' is learned.

Let us examine these facts. That is, let us examine the truth conditions Smart offers for propositions ascribing rule following, and in particular for the proposition ascribing the green-following rule as opposed to the grue-following one. It is only fair to preface this by pointing out that there is no innovation in proposing to turn to dispositions in order to solve the problem of rule ascription. Smart's innovation has to do with his acceptance of Wittgenstein's claim that there is no room for ascribing to someone the following of a rule in the absence of a clear notion of what a correct application of it, as opposed to its misapplication, consists of. This requirement seems to be that the ascription of rule following be couched entirely in normative terms, and not within the framework of an explanatory scientific theory. This is where Smart innovates. He puts forward the claim that there is an empirical sense to a *correct* application of a rule in dispositional terms. The application is correct if it is in accordance with the disposition and without random interference, internal or external. This *is* an innovation, which will have to be scrutinized.

One fact in virtue of which I can, according to Smart, truthfully say that I am using 'green' in the sense of green and not of grue is my physiology. My physiological structure, according to this argument, is capable of supporting the counterfactual proposition that, had I lived to and beyond the year A.D. 2500, I would have continued to apply the word 'green' to the color of the grass and refrained from applying it to the color of the sky. The categorical base of a disposition provides a firm enough basis to ascribe to me the following of a rule, even if I had as a matter of fact applied this rule just once in my life.

This way of solving the problem, however, cannot in my opinion answer the central difficulty, which is, what would count as a correct application of 'green' after the year A.D. 2500. To be sure, Smart's answer would be that the correct application is the one that accords with the disposition whose base is in no way interfered with. But then, one queries, is not just such an interference necessary in order to assure that I shall still be alive until the year A.D. 2500? After all, we lose brain cells constantly, and in order for me to live like a Methuselah without my senility causing me to apply 'green' to the color of the sky, some rather massive intervention is called for in my brain. But then, of course, the entire story about the categorical base of my disposition loses its power, for my brain will then be very different from my present brain. In other words, we do not really know what the conditions are under which all other things are equal. And these precisely are the conditions which, according to Smart, are supposed to tell the correct application apart from the misapplication of 'green'. The lesson to be drawn from this, in my view, is that the ceteris paribus conditions are no substitute for the normativity of a rule — not even in an empirical, scientific theory.

This difficulty was presented by Kripke in Wittgenstein's name. And even if it is not really a problem of Wittgenstein, it remains a difficulty nevertheless.

As I have already mentioned, Smart has, in addition to the disposition argument, another argument about facts by virtue of which I am indeed a green speaker, not a grue speaker. It is an argument addressed specifically at the grue-green example, and has to do with the way we come to learn our adjectives. Since we all learn our color terms by ostension, it is clear, according to Smart, that the way of learning 'green' is different from the way of learning 'grue'. Thus, in the case of learning 'grue' at least two color samples are required (namely, green and blue), whereas only one is required in the case of learning 'green'. Moreover, the process of learning 'grue' essentially involves some reference to dates, over and above the ostension, which is not the case with 'green'. True, Wittgenstein went a long way toward making us aware of the elaborate stage-setting often required for ostensive definitions. But, whatever the setting, there is still, according to Smart, an asymmetry between the ostensive way of coming to learn 'green' in the sense of green and in the sense of grue.

My reply would be that it is only by begging the question that one gets this asymmetry. My claim, in other words, is that the way of coming to learn 'green' and 'blue' is identical to the way of coming to learn 'grue' and 'bleen'. It is only when you somehow happen to *know* that you are a green speaker, and then proceed to ask how you are to learn the new color term 'grue', that you get the asymmetry Smart talks about. But when the issue is considered without begging the question i.e., is there asymmetry between the ways of coming to learn 'green' as opposed to 'grue' from the ostension sample comprising, say, emeralds, lawns, and the eyes of Queen Surayah, then the answer must, I contend, be negative. It is an error to hold that only 'green' involves *direct* ostension, that is, that the term introduced by the ostension truthfully applies to the surface indicated by the finger, whereas 'grue' involves *deferred* ostension — deferred, that is, to that blue beyond the year A.D. 2500. The situation rather, is that in both cases we point at the emerald and in both cases we truthfully apply 'green', whether in the sense of green or in the sense of grue. Whatever it is that looks puzzling with regard to the learning of 'grue' and 'bleen', from the point of view of green and blue, is every bit as puzzling with regard to the latter terms, when considered from the opposite point of view. The symmetry regarding learning is complete.

The conclusion I draw, then, is that neither disposition nor ostension provide the elusive truth condition for my using 'green' in the sense of green and not grue.

A retort common on occasions such as this is: how do you know that there is a difference between 'green' in the sense of green and in the sense of 'grue' anyway, given that you do not know in which sense you are using the term?

An analogy might help. I sometimes use the terms 'stalagmite' and 'stalactite' without knowing which of them refers to the structure forming from below and which from above. I can still, of course, talk of them as different, or contrasting. Similarly, I can be said to know that there is a difference between 'green' and 'grue', and that their future applications diverge, without at the same time knowing whether I am pointing out the difference as a green-blue speaker or as a grue-bleen speaker.

So much for the issue of truth conditions to propositions ascribing rule following. And now we must, as promised, turn to the interpretation issue.

There are three different, though related, contexts within which the question of meaning might be dealt with. First, the learning context, namely, the way a term or expression is learned for the first time. This is the context in which Smart advances his claim that the learning of 'green' is different from the learning of 'grue'. Second, the explanatory context, where one deals empirically with the principles constituting the mechanism that enables us to use the language. This is the context in which Smart has accounted for rules through the mechanism responsible for our linguistic dispositions. And finally, there is the normative context, where one is concerned with criteria for the resolution of disputes about meaning, as well as with criteria for guiding our linguistic behavior.

It seems that Wittgenstein limited the philosophical discussion to the last of these contexts, that is, to the normative one. True, Wittgenstein offers many thought experiments regarding the way various expressions are learned; so much so, indeed, that at one point he asks himself in alarm whether he is turning into a child psychologist. However, Wittgenstein's concern with the context of learning on these occasions is for one single purpose only. It is to extricate from the explanations of meaning given to the learner, whether a child or a foreigner, the meaning rules thereby conveyed. That is, the learning context is of interest to Wittgenstein inasmuch as one can find there normative indications as to how to use a given expression correctly. It is thus a meaning explanation he is looking for in the learning context, not an explanation of the learning itself. As far as Wittgenstein himself is concerned, one might conjecture, it is conceptually possible for a language not to be learned at all but rather to be innate.

The explanatory context, too, is not for Wittgenstein a legitimate one for philosophical discussion, since it is all conjectural. The province of philosophy, for Wittgenstein, is criteria, not conjectures.

The fact that Wittgenstein, in contrast to Smart, does not see philosophy as a continuation of science by other means is no real reason for us to disregard Smart's position, according to which the explanatory and the normative contexts are complementary rather than competitive, and the division of labor between philosophy and science is not well delineated. However, I take Wittgenstein to have believed also that there is no room for a systematic account of meanings. Put differently, language for Wittgenstein has no "essence," it has no uniform character. For him the idea that a uniform and systematic account can be given to language, and that there is therefore a single method for endowing the expressions of the language with meanings, is an illusion. This illusion may, according to Wittgenstein, be formed from the impression created by the written language. The written language appears uniform and susceptible to systematic description. But this is misleading, much like the impression created by the keyboard of a complicated machine. The similarity of the keys may lead one to expect similarity of function where no similarity exists.

For my part, I believe that Smart is right: there is indeed room for the explanatory context as complementing the normative one. (I even deal with this issue at some length elsewhere.) However, in contrast to Smart, I do not believe that rules of meaning can be accounted for within a purely descriptive theory, without a normative element. Analogously, one cannot account for the psychology of mathematical thinking without normative mathematics playing a crucial role in the description, whether as adhered to or as deviated from.

George Bernard Shaw's golden rule that there are no golden rules holds firm. There is no rule that can rule out extreme interpretations. But then we follow rules blindly and not through interpretation. Smart has offered an explanation for what counts as following a rule blindly; he has not, however, accounted for what counts as following a rule bindingly.

Why Does Physics Need Mathematics?

JEAN-MARC LÉVY-LEBLOND

1. The Effectiveness of Mathematics in Physics

The existence of a specific and peculiar relationship between physics and mathematics is an obvious and often-recognized fact. Many explicit testimonies occur throughout the history of physics, starting with the famous statement by Galileo:

> Philosophy [Galileo is obviously referring to natural philosophy] is written in that great book permanently open to our eyes — I mean the Universe — but one cannot understand it if one does not first learn to know the language and the characters in which it is written. It is written in mathematical language, and its characters are triangles, circles, and other geometrical figures, without which it is humanly impossible to understand a single word of it. (Galileo 1623, 231.)

Three centuries later, the astrophysicist J. Jeans writes:

> What we are finding, in a whole torrent of surprising new knowledge, is that the way which explains [it] more clearly, more fully and more naturally than any other is the mathematical way, the explanation in terms of mathematical concepts.... From the intrinsic evidence of his creation, the Great Architect of the Universe now begins to appear as a pure mathematician. (Jeans 1973, 153, 167.)

One could assemble a full anthology of such quotations, and indeed any chapter of a physics textbook will provide examples. What would classical mechanics be without differential and integral calculus; electromagnetism

E. Ullmann-Margalit (ed.) The Scientific Enterprise, 145–161.

without partial differential equations; general relativity without tensor calculus; quantum theory without linear algebra; and so on?

It is clear, then, that physics successfully "uses" mathematics; and even though this statement, as we shall see, is not purely factual and involves certain presuppositions, it certainly renders a fair account of the apparent situation. But it immediately leads one to wonder about the reasons and causes of this success. How is it that mathematics, the study of purely abstract concepts and structures, works so well in helping physics to describe and explain the concrete and complex material world? That this very adequacy presents a problem is often acknowledged by physicists, with ingenuous surprise or in an embarrassed vein:

> It is nevertheless ... remarkable that, among the abstract constructions elaborated by mathematicians, guided exclusively by their need for logical perfection and increasing generality, none seems to remain useless for the physicist. By a peculiar harmony, the needs of the mind, eager to construct an adequate representation of reality, seem to have been foreseen and preceded by the logical analysis and abstract aesthetics of the mathematician. (Langevin 1950, 88.)

> Thus, with each chapter of mathematics, we have at our disposal an infinity of possible games.... We have assumed that it is possible to select some of these games which can be put in correspondence with all or part of physical reality.... Such success — thanks to which mathematical physics exists — could not have been anticipated a priori. (Vogel 1956, 14.)

> The idea that mathematics could so to speak be applied to the objects of our experiences seemed to me remarkable and fascinating. (Heisenberg 1971.)

Finally, one could quote the well-known title of an essay by E. Wigner, "The Unreasonable Effectiveness of Mathematics in the Natural Sciences" (Wigner 1967), in which the author is in fact dealing specifically with physics.

2. Mathematics — the Language of Physics?

Various solutions have been proposed to the problem of this connection between physics and mathematics. Whether made by scientists or philosophers, the overwhelming majority of these suggestions rely on the idea that mathematics constitutes the language of physics. In addition to the already quoted text of Galileo, we cite only three representative opinions:

> [Physical] laws are always drawn from experience, but in order to state them one requires a special language; ordinary language is too poor, also too vague, to express relationships so delicate, so rich, and so precise. This, then, is the first reason why the physicist cannot do without mathematics: it furnishes him with the only language he can speak. (Poincaré 1913, 180.)

> Mathematics so to speak constitutes the language with which a question may be asked and answered. (Heisenberg 1971.)

> Mathematics (first) provides the language in which the quantitative predictions of experiments are expressed. (Wightman 1976, 16.)

However, this picture of mathematics as the language of physics is rather ambiguous, as it admits two opposing interpretations: is mathematics the intrinsic language of Nature, which Man must master in order to understand her workings; or, conversely, is it a special language of Man into which the facts of Nature must be translated in order to be apprehended? The first position seems to be Galileo's; it is also the view of Einstein:

> According to our experience up to now, we may rightly be convinced that Nature is the realization of the simplest mathematical ideas. I am persuaded that pure mathematical reasoning enables us to retrieve those concepts, and the principles relating to them, which gives us the key of natural phenomena. (Einstein 1934.)

The second point of view is stated explicitly by Heisenberg:

> Mathematical formulas no longer represent Nature, but only our knowledge of it. (Heisenberg 1971.)

It should be stressed, that these two attitudes, far from being contradictory, are simply two extreme positions on a continuous spectrum of opinions. In addition to the thematic couple Nature-Man, other conceptual pairs — such as experiment-theory, concrete-abstract, facts-laws — define the common framework of such theories, ranging from empiricism to formalism.

However, the conception of mathematics as a language does not answer our original question: it misses the goal by overshooting it. Indeed, instead of explaining why mathematics applies so specifically to physics, it provides us with an overly vague solution; it is concerned with the supposed adequacy of mathematics in respect to scientific knowledge in general, to the study of nature in its totality. Despite the apparent differences, the various views of mathematics as a language share a general conception, according to which

Science, over and above the specific sciences, finds in mathematics a single, universal method of representation and analysis (it is worth noting — though this is beyond the scope of the present investigation — that those who profess such views usually hold that there is in Science a single and universal experimental method). But we are interested here in the peculiar relationship of mathematics to physics, so that nothing much has been achieved. Let us try, then, to start anew: instead of asking "*Why* ...?" let us first ask "*How* does mathematics apply to physics?"; in other words, before inquiring as to the foundations of the relationship of mathematics to physics, let us inquire about its nature.

3. The Nature of the Relationship of Mathematics to Physics

We start by observing that the usual statement "mathematics applies to other sciences," although seemingly innocent, already implies a strong standpoint as to the crux of the matter, since it characterizes the relationship of mathematics to these other domains as one of *application*. The implication is, that we are concerned with an instrumental relation, in which mathematics appears as a purely technical tool, external to its domain of implementation and independent of the concepts proper to that domain. Such a description seems justified enough in the cases of chemistry, biology, the earth sciences, etc. — i.e., in general, for most natural sciences, except physics. In these (nonphysical) sciences, the role of mathematics is reduced mostly to numerical computations, to quantitative reckoning. Applications may range from the entirely elementary, such as the trivial arithmetical balancing of chemical reactions, to the highly sophisticated, such as the use of statistical methods in genetics. Nevertheless, all such situations are characterized by a clear-cut separation between the conceptual equipment proper to the considered domain on the one hand, and the mathematical techniques used, on the other. More precisely, and to select some more or less arbitrary examples, consider such fundamental concepts as those of pure and simple bodies; chemical binding, reduction and oxydation in chemistry; primary, secondary, and tertiary structure of proteins, enzymatic activity, allostery, the genetic code in biochemistry; sedimentation, facies, geosyncline in geology; these concepts clearly involve nothing mathematical, whether in their definitions or in their operation.

The situation is completely different in physics, where mathematics plays a much deeper role. Indeed, it would be almost impossible to find a physical concept which is not intimately and indissolubly linked to one or several mathematical concepts. For instance, how could one think efficiently about the concept of velocity without the concept of derivative? How could one think about the electromagnetic field without vector fields and vector calculus? How could one ponder the principle of relativity without group theory? How could one consider quantum-physical magnitudes without self-adjoint operators in Hilbert space? Mathematics is thus internalized by physics. We shall say that the relationship of mathematics to physics is one of *constitutivity*. A similar idea has been expressed by Bachelard:

> The hypotheses of physics are formulated mathematically. They become inseparable from their mathematical form: they are really mathematical thoughts.... One must break with the stereotype of the skeptical philosophers for whom mathematics is but a language. On the contrary, mathematics is *a way of thinking* — a way of thinking confident in its language. The physicist experiences nature through this way of thinking.... Mathematism is not descriptive, but constitutive. (Bachelard 1965, 29.)

In order to further clarify the meaning of this constitutive relationship, we emphasize that a physical concept cannot be identified with or reduced to its underlying mathematical concept(s). Neither is a physical concept a mathematical one plus a physical interpretation: the mathematical concept is not a skeleton fleshed out by physics, or an empty abstract form to be filled with concrete content by physics. The distinction between a physical concept and its mathematical characterization cannot be thought of in static terms. Rather, mathematics, as Bachelard puts it, is a dynamical "way of thinking," leading from the experimental facts to the constitution of the physical concept. The history of the concept of velocity could be used as a typical example, starting with Galileo and his difficult elaboration of the idea of instantaneous speed via Euclid's theory of proportions (Ravetz 1972, 11), continuing with Newton's clarification using his new calculus, up to relativistic developments from Einstein to Minkowsky, and ending with the modern distinction between velocity and "rapidity."[1]

Even more interesting as counterproofs are the many attempts, always stillborn but nevertheless reborn, to "demathematize" physics. Whenever physics undergoes a new phase of abstract mathematization, a nostalgic rearguard — which does not consist solely of the unavoidable crackpots and cranks — demands a return to a "more intuitive" and "less mathematical" physics. Long before such criticism was aimed at relativity and quantum theory in this century, it was leveled against Maxwell's field electromagnetism and Newton's gravitation (see, for instance, the most interesting and nontrivial corpuscular gravitation theory of Lesage, which tried to "explain away" action-at-a-distance[2]). Such negative judgments in fact correspond to conventional epistemological viewpoints, according to which the mathematics/physics relationship is assimilated into an abstract/concrete one. From such a standpoint, it is asserted that at each new step of physics its laws become "more and more mathematical," and its concepts "more and more abstract." Orthodox physicists will accept the situation as a remarkable conquest or as a necessary evil; heterodox physicists and "outlaws" will reject it; but they will all agree on the diagnosis. Still, can one really say that, intrinsically speaking, a derivative is "more mathematical" than a ratio, a partial differential equation "more mathematical" than a differential equation, a complex number "more mathematical" than a real one, a Hilbert space "more mathematical" than a three-dimensional Euclidean space, and so on? Is it not clear that this is a question of historical contingencies, and that the degree of conceptual abstraction and mathematical sophistication of a physical theory, linked as it is to a given place and time, cannot be used as an epistemological criterion?[3] A more detailed examination of the constitutive relationship of mathematics to physics is thus necessary.

4. The Mathematical Polymorphism of Physics

We first explicitly reject the rather Platonic interpretation of the connection between mathematics and physics, according to which the role of the physicist is to decipher Nature so as to reveal "the hidden harmony of things" (Poincaré), as expressed by mathematical formulas, underlying the untidy complexity of physical facts. We reiterate that to speak of a constitutive relationship is not to say that every physical concept is the contingent materialization of an absolute mathematical being which is supposed to express its deepest truth, its ultimate essence. To illustrate the dynamical nature of the relationship, it suffices to consider an essential characteristic of physical concepts and laws, namely, their *mathematical polymorphism*: the possibility that each and every one of these laws and concepts may be endowed with several mathematizations. An elementary example is uniform rectilinear motion, which may be thought of in geometrical terms: equal spatial distances covered in equal times (Galileo); in functional terms: linear dependence of the distance on time; or in analytical (differential) terms: constant velocity or zero acceleration (Newton). A less trivial example is the classical dynamics of point masses which may be formulated in terms of differential equations (Newton), partial differential equations (Hamiltonian viewpoint), variational equations (Lagrangian viewpoint), etc. Of course, such different formulations are strictly equivalent, mathematically and logically speaking. But they are not so, physically speaking; there are clear, though subtle, distinctions between them, since the advent of a new formulation usually corresponds to the necessity of solving new problems or to the impact of new ideas (originating either in other branches of physics or in mutations within mathematics).

In some cases, the coexistence at a given time of various formulations of a law or concept reflects the persistence of archaic elements in a domain which has not yet been subjected to an epistemological recasting — curiously, this is the situation today in such "modern" domains of physics as relativity or quantum theory (Lévy-Leblond 1976a, 617; 1979, 23; 1977a, 171; 1977b, 415). In general, however, this polymorphism rather expresses the diversity of situations in which a given law must be implemented and its more or less efficient working, depending on which formulation is used to deal with the specifics of the situation. To cite a previous example, classical dynamics will be used mainly in its Newtonian form if one is interested in computing trajectories, but its Hamiltonian form will be preferred instead if one is looking for conservation laws. Similarly, the choice between local formulations of electromagnetism (differential field equations) and global expressions (flux or circulation theorems, such as those of Gauss or Ampère) is dictated by the circumstances of the problem at hand. This is the reason why physics, in contrast to mathematics in its modern form, is not readily amenable to axiomatization. Given a set of different statements, all of which can be deduced from one another, the physicist hesitates to impose a hierarchical ordering. "Principles," "laws," and "applications" in physics seem to possess a relative mobility, an interchangeability, to a much higher degree than axioms, theorems, and corollaries in mathematics. (See the excellent discussion in Feynman 1967.)

These remarks are even more pertinent with respect to the future of physics than to its past. Among the various equivalent formulations of a given physical law available at a certain time, only one or a few of them, in general, will extend most naturally to new phenomena. Such extensions will open the field of relevance of these "favorable" formulations, while restricting and limiting the domain of the others; this discrimination violates the (physical) equivalence. To take anew the same example, Einsteinian relativistic mechanics hardly tolerates a Newtonian formulation (based as it is on the idea of instantaneous action-at-a-distance), while the Lagrangian formulation is most natural (even more so than in the nonrelativistic case).

Of course, the importance of a specific formulation cannot be attributed solely to its intrinsic, theoretical merits. Every one of these formulations carries with it a complex set of mental representations; it is associated with specific world views and is preferred by certain schools of thought. This "ideological matrix" may play an essential role in physicists' adoption or rejection of the considered formulation, at the individual or collective level. As an example, it is most interesting to study in detail the teleological and theological implications of variational principles, from Maupertuis to Feynman (see the historical and epistemological discussions in Yourgrau and Mandelstam 1968). Conversely, a physicist's philosophical or metaphysical preferences may well condition his privileged attachment to such or such a type of formulation and, according to the case, may propel or impede his work; it may be held, for instance, that Einstein's predilection for geometrized physics helped him discover general relativity, while obstructing his acceptance of quantum theory.

Recognition of the intrinsic mathematical polymorphism of physics obviously has deep epistemological and educational consequences — but this is another matter.

5. The Multivalence of Mathematics in Physics

In order finally to dispel the tempting illusion of a preestablished harmony between physical and mathematical concepts, we draw attention to the physical *multivalence* of mathematical structures (which is the converse of the mathematical polymorphism of physical laws), i.e., to the possibility that the same mathematical theory may be adequate to several completely different physical domains. Indeed, a well-known example is the common description using second-order linear differential equations (with constant coefficients) of mechanical vibrations and electrical oscillations — and many other phenomena. Similarly, Poisson's partial differential equation governs electrostatics, Newtonian gravitation, heat diffusion, (stationary) neutron diffusion, elastic membrane equilibrium, laminar flow of a two-dimensional fluid, etc. Of course, there is no need for an idealistic interpretation of these formal identities, which have been beautifully explained by Feynman:

> However, a question surely suggests itself at the end of such a discussion: why are the equations from different phenomena so similar? We might say, "It is the underlying unity of nature." But what does that mean? What could such a statement mean? It could mean simply that the equations are similar for different phenomena; but then, of course, we have given no explanation. The "underlying unity" might mean that everything is made out of the same stuff, and therefore obeys the same equations. That sounds like a good explanation, but let us think. The electrostatic potential, the diffusion of neutrons, heat flow — are we

> really dealing with the same stuff? Can we really imagine that the electrostatic potential is physically identical to the temperature, or to the density of particles? Certainly it is not exactly the same as the thermal energy of particles. The displacement of a membrane is certainly not like a temperature. Why, then, is there "an underlying unity"?
>
> A closer look at the physics of the various subjects shows, in fact, that the equations are not really identical. The equation we found for neutron diffusion is only an approximation that is good when the distance over which we are looking is large compared with the mean free path. If we look closely, we would see the individual neutrons running around. Certainly the motion of an individual neutron is a completely different thing from the smooth variation we get from solving the differential equation. The differential equation is an approximation, because we assume that the neutrons are smoothly distributed in space.
>
> Is it possible that this is the clue? That the thing which is common to all the phenomena is the space, the framework into which the physics is put? As long as things are reasonably smooth in space, then the important things that will be involved will be rates of change of quantities with position in space. That is why we always get an equation with a gradient. The derivatives must appear in the form of a gradient or a divergence; because the laws of physics are independent of direction, they must be expressible in vector form. The equations of electrostatics are the simplest vector equations that one can get which involve only the spatial derivatives of quantities. Any other simple problem — or simplification of a complicated problem — must look like electrostatics. What is common to all our problems is that they involve space and that we have imitated what is actually a complicated phenomenon by a simple differential equation.(Feynman et al. 1963, 2:12f.)

These remarks of Feynman are obviously of a general nature and are valid for all cases of multivalence. It is through a common process of approximation/abstraction that different physical phenomena are encompassed by analogous mathematizations. We note, further, that such convergences lead to the construction of new "multivalent" physical concepts, with operational validity in various domains; consider, for instance, the notions of resonance, impedance, etc., as used in all (linear) oscillating phenomena, whether they are mechanical, acoustical, electrical, etc.[4] It is quite clear that such physical concepts are indeed based on and constituted by a specific mathematical formulation (in the above example, the theory of linear second-order differential equations with constant coefficients, which are endowed with a physical meaning through generic identification of the various terms as inertial, dissipative, restoring, and exciting "forces").

The multivalence of mathematics in physics is closely related to the internal multivalence of mathematics, that is, to the emergence of general notions and theories with diversified realizations. Clearly, the idea that mathematical structures exist antecedently to the physical concepts they underlie is neither ontologically valid, as we have seen, nor historically so. Many essential mathematical theories were born from the very trend toward abstraction and generalization which revealed their physical multivalence; consider, for instance, the emergence of integral and differential calculus from classical mechanics, or, much later, the birth of differential geometry from field theory. This inverse relationship, from physics to mathematics, deserves a full study of its own. One would probably discern a dual and contradictory evolution within mathematics by which, on the one hand, mathematics has become increasingly independent, with its own ways of development, while, on the other hand, it continues to find motivations and resources in physics. There are indeed some interesting cases of modern advances in mathematics that stem from physical problems, such as the theory of locally compact groups and harmonic analysis, the theory of dynamical systems, etc. However, it must be emphasized that, as mathematics develops on its own, that fraction of its theories and ideas utilized by physics is decreasing rather rapidly. While it is true that "physical mathematics" keeps on growing, it certainly does so much less rapidly than "mathematical mathematics." In short, and contrary to some rash assertions (some of which were cited above), it is not true (and is becoming ever less so) that "none (of the abstract constructions elaborated by mathematicians) seems to remain useless for the physicist" (Langevin 1950, 88). There is no need, then, to marvel at the mysterious harmony and correspondence between mathematical structures and physical phenomena.

6. Mathematical Physics

It seems appropriate at the present stage to dwell briefly on the peculiar status of "mathematical physics" as the specialized branch of physics. Mathematical physics should be distinguished from theoretical physics in general. The latter expresses and applies physical laws, creates physical concepts, and puts them to work, under the constraint of experimental physics and in close interaction with it. Theoretical physics exhibits different levels, ranging from the interpretation of quite specific experimental results in terms of well-known physical laws to the search for (and, hopefully, the discovery of) new laws. It is mathematical in the sense that mathematics plays in it the constitutive role that we have already described.

Mathematical physics, as such, is a far more specialized activity, consisting mainly in the recasting and refining of theoretical physics. In fact, physical theories are necessarily branded at birth by the historical circumstances of their production. Their formulation often bears traces of the intellectual struggles accompanying their discovery, at the very least through the terminology used; moreover, their formalism is linked to a certain, chronologically determined, state of the mathematical art. Physical theories thus look like buildings, more or less recently completed, but still covered with scaffolding and strewn with the remains of previous constructions. Removing the scaffolding, getting rid of the remains, fully revealing the intrinsic structure of the building, assessing the depth and solidity of its foundations, as well as discovering its weak points — this is the main task of mathematical physics. It enables one to evaluate the precise degree of correlation among the different parts of the theoretical structure, and to estimate its rigidity or softness, i.e., its ability to withstand necessary or potential changes (see, e.g., Wightman 1976).

As an example, let us consider certain developments in special relativity. The core of this theory, i.e., the invariance of physical laws under the Lorentz transformation group, was established by Einstein, who relied, among other assumptions, on the hypothesis that the velocity of light is constant (invariant). This constancy, however, is merely an experimental result, of high — but limited — precision, which could be refuted — for instance, if the photon had a nonzero (however small) mass. Were this to be the case, would relativity theory itself collapse? Here, mathematical physics is capable of establishing the nature of the link between the theory and its basic assumption. In the present case, it is known today that the constancy of the velocity of light is a sufficient, but not necessary, condition for the validity of special relativity. The theory may be based on more general and more secure hypotheses, starting from the abstract principle of relativity and using group theory (see Lee and Kalotas 1975, 434; Lévy-Leblond 1976b, 271; Lévy-Leblond and Provost 1979, 1045 and additional references therein). In fact, today one interprets the constancy of the velocity of light as a special consequence of Einsteinian relativity, due to the (apparently) vanishing mass of the photon.

Here, again, we encounter the difficulty — and the danger — of a canonical axiomatization of physics. Indeed, a given law or theory may be deduced from several different sets of principles, more or less general, more or less plausible. If an experiment contradicts the theory then, obviously, one basic principle at least in each of these sets is wrong. But, conversely, the falsification of one such principle in a given set does not necessarily imply failure of the theory.[5] However, mathematical physics is not the province of logicians or pure mathematicians, for the very nature of the theories it is investigating at any given time, and the type of alternative hypotheses it is led to propose, must correspond to the general development of physics and take its experimental aspects into account. The limits confronting such a great mathematician as Poincaré in his work in physics, especially in regard to relativity, clearly illustrate these remarks.

Finally, we wish to note that, if a specific part of theoretical physics can be classified as mathematical physics, this does not necessarily imply a clear-cut and institutionalized separation. The existence of a well-defined branch of mathematical physics depends upon the domain of physics and the period in question. The very term "mathematical physics" took hold, apparently, at the end of the last century, when classical physics was at its zenith; it then fell more or less into oblivion during the period of the rapid development of modern physics, to surface anew in the 1950s. This reappearance was probably due to the considerable technical difficulties that arose in certain new domains (such as the quantum theory of fields), but also to the increasing specialization of scientists.[6]

Just for the sake of completeness, we should also mention another aspect of mathematical physics, of a more ancillary and less fundamental nature, relating to technical rather than epistemological problems. We are referring to the constantly ongoing work of refining the specific mathematical tools of physics, at a level much too trivial for the mathematician, but most useful for the physicist: extending the repertory of the special functions of physics, the catalog of known differential equations, tables of integrals, lists of classified groups, and so on.[7]

7. Mathematics and the Specificity of Physics

We can now return to our main problem. The singularity of physics in its relationship to mathematics, as sketched in the preceding sections, is clearly very difficult to account for on the assumption that mathematics is merely a language. Indeed, within such a framework, this "language" would necessarily have to be a universal one, applicable in principle to each and every natural science. The specificity of physics — a field in which it is empirically obvious that mathematics "works better" — must then be treated as only a matter of degree, and criteria must be found to distinguish it from other sciences.

A possible (and common) view is to consider physics as "more advanced," which would explain its more developed mathematization. It is argued that more precise experimental methods, a firmer control of experimental conditions, permit the quantative measurement of physical notions, which are thus transformed into numerical magnitudes. This generalized measurability would then explain the intervention of mathematics, "the science of numbers." But this argument per se still fails to explain the privileged status of physics; moreover, it relies on a very naive view of mathematics. To reduce mathematics to the handling of numbers is as shortsighted as to consider it as a language. Even those branches of mathematics most usually employed in common numerical calculations, such as elementary function theory, involve fundamental concepts which are not of a numerical nature, such as the derivative. A fortiori, the role of abstract group theory or functional analysis in quantum physics, for instance, is at most remotely connected with the quantitative aspects of physical measurements.

One might be tempted instead to localize the specificity of physics in its objects of inquiry rather than in its being (allegedly) historically more advanced. It is commonly stated that physics is "more fundamental" than the other natural sciences. Dealing with the deepest structures of the world, it aims to bring to light the most general laws of Nature, which are implicitly considered as the "simplest" and thus, in an Aristotelian way, as the "most mathematical." This view directly implies a hierarchical ordering of the sciences, with physics as the "queen of sciences," as has been claimed. Thus mathematicity assumes a normative character and becomes a criterion of scientificity. This is, of course, quite a classical standpoint:

> No human investigation can be considered to be proper science if it does not go through mathematical proofs. (Leonardo da Vinci, 41.)

> I hold that in any specific natural science there can be found only so much science proper as there is mathematics present in it. (Kant 1786, Preface.)

However, the very development of the various sciences flatly contradicts this point of view, whether we consider the persistence of "old" sciences, such as chemistry or geology, as autonomous domains, or the emergence of new sciences, such as molecular biology. We have already stressed that these sciences have their own, nonmathematical, concepts, whose mutual consistency and correspondence with the specific experimental practices of the domain are quite sufficient to ensure their respectable scientificity according to the criteria of most philosophers — including, in any case, all concerned scientists. As a matter of fact, the knowledge of atomic structure offered by modern physics and the ensuing possibility of a detailed understanding of chemical bonds or reactions do not render the notions of chemistry obsolete or useless; quite on the contrary — by revealing the very limits of these notions, the physical reduction makes them more profound. Furthermore, the reduction is only one of principle, which, save in a few simple cases, cannot replace the specific level of the chemical theory: to solve the Schrödinger equation for eighteen electrons in a field of three nuclei with charges +1, +1, and +16 and unknown positions, is neither the simplest nor the most efficient way to understand the properties of the water molecule! In other words, when progress in a scientific field affords access to a domain hitherto dependent on another field, it is not a question of simply displacing the borderline to the benefit of the newcomer and to the detriment of the previous one. Rather, one is faced with a status of dual nationality, with all the inherent advantages and disadvantages.

To sum up our discussion, we seem to face a double difficulty: on the one hand, we must explain the singular relationship of mathematics and physics; and on the other hand, we have to understand the specific distinction between physics and the other natural sciences. Neither the history, the subject matter, nor the experimental methods of physics provide an answer to both questions: any attempt to solve one of them at once deepens the mystery of the other. We are thus led to propose a radical alternative, in which both questions are answered simultaneously — or, rather, mutually "canceled out." Our Columbus-egg-type thesis is the following: the relationship of mathematics and physics *is* the specific determination of physics. In other terms, it is through the role that mathematics plays in a given area of the natural sciences, according to its constitutive (or otherwise) character, that this area is recognized (or not) as belonging to the domain of physics.

There are strong historical arguments to back up this point of view. After all, it is surely the role of mathematics in Galileo's work that characterizes the "great divide," marking the birth of physics in the modern sense of the word. This was explicitly recognized by Descartes, writing in a letter to Mersenne (11 October 1638):

> I consider that he [Galileo] philosophizes much better than the common herd, in that he abandons the errors of the School, and tries to elucidate physical matters through mathematical reasoning. In that respect, I entirely agree with him and I hold that there is no other way to reach the truth. (Descartes 1952, 1024.)

Descartes is referring here to the book which Galileo himself quotes in his correspondence as his *Treatise on Motion*; however, the title under which this book was published, probably because the publisher wanted to draw attention to its novelty, was *Mathematical Discourses and Demonstrations Belonging to Two New Sciences*.[8]

A century and a half later, electricity and magnetism entered physics proper only through the mathematical formulation of Coulomb's law. Previously, they were intimately bound up with vitalist conceptions and the study of life phenomena, which, strange as it may seem today, did not prevent notable experimental advances (Leyden jars, electrostatic machines, voltaic cells). The radical novelty of the "mathematical" point of view was clearly expressed in 1810 by Delambre, writing in a progress report to the French Academy of Sciences:

> Every problem concerned with light, gravity, motion, and collisions of bodies, is today ruled almost exclusively by geometry.... There has *even* been an attempt to submit the phenomena of electricity and magnetism to calculus.

Another example would be the intervention of mathematics, through groups theory, that brought crystallography from the earth sciences into physics at the end of the nineteenth century (Maitte 1984).

From an epistemological vantage point, it must be recognized that borderline fields such as physical chemistry, geophysics, biophysics, etc., cannot be defined either by their objects of investigation — which are obviously the same as those of chemistry, geology, and biology, respectively — or by their experimental methods, which are much too diversified and rapidly evolving. Rather, it may be maintained that it is the nature of their specific concepts, mathematically constructed. A recent example is furnished by the new framework of geophysics, plate tectonics, in which a seminal role was played by geometrical considerations about motions on a spherical surface (Euler's theorem; see Hallam 1973, esp. chap. 6). Finally, let us consider the case of astrophysics, which is most interesting in that it enables us to eliminate one possible misunderstanding of our main thesis. Indeed, one must be careful not to adopt a formalist interpretation, according to which the specific (constitutive) relationship between physics and mathematics would be considered as subordinating the former to the latter. For one must not "forget" the existence of experimental practices, which set apart the natural sciences — including physics — from mathematics. Astronomy, a much older science than physics, was long considered a part of mathematics, even well after the birth of physics. The foundation of astrophysics at the end of the nineteenth century, and its progressive domination of classical astronomy, resulted precisely from the import of such experimental practices in an already mathematically constituted domain, which produced a new and active branch of physics proper. This example shows that displacement of frontiers does not necessarily occur in a single direction within a positivist classification of the sciences. As we have tried to show in the case of physics and mathematics, the very idea of a universal classification and of hierarchical relationships is apt to obscure the necessity of understanding the specificity of the various sciences and the peculiarities of their relationships.

Notes

1. For a didactical presentation of the contemporary views of theoretical physics, see Lévy-Leblond 1980, 345.
2. Lesage's theory is presented by S. Aronson (1964, 51). For a brief discussion in the more general framework of heterodoxy in science, see Lévy-Leblond 1982, 353.
3. An amusing example is given by P. Langevin in his recollections of his undergraduate course in physics, in which the introduction of electrical voltage (!) was considered a bold innovation, concerning an elaborate and immaterial notion. But then, of course, there were no sockets in the walls of one's home, delivering 220V (Langevin 1950, 115).
4. See, for example, the recent interesting textbook based on the commonality of such ideas — and of the corresponding devices — Shive and Weber 1982.
5. We are close here to Duhem's argument.
6. It is interesting to note that the most important specialized periodicals in the domain today, the *Journal of Mathematical Physics* and *Communications in Mathematical Physics*, were established in 1960 and 1965 respectively, that is, rather recently.
7. An elementary but amusing example is provided by the recent proposals and discussions of new simple numerical approximations of the factorial function, supposed to improve on the classical, centuries-old, Stirling formula. See Weissmann 1983, 1985.
8. See the introduction by A. Favaro to Galileo's *Two New Sciences.*

References

Aronson, S., 1964. *Natural Philosopher* 3:51.

Bachelard, G., 1965. *L'activité rationaliste de la physique contemporaine.* Paris: Presses Universitaires de France.

Descartes, R., 1952. *Oeuvres et Lettres.* Paris: La Pléiade.

Einstein, A., 1934. *The World As I See It.* Lyle Stuart.

Feynman, R. P., 1967. *The Character of Physical Law.* Cambridge, Mass.: MIT Press.

Feynman et al., 1963. *The Feynman Lectures in Physics*, Vol. 2. Reading, Mass.: Addison-Wesley.

Galileo Galilei, 1623. *Il Saggiatore.* Vol. 6. : Ed. Naz.

—, 1914. *Mathematical Discourses and Demonstrations Belonging to Two New Sciences*. New York: Dover.
Hallam, A., 1973. *A Revolution in the Earth Sciences*. Oxford: Oxford University Press.
Heisenberg, H., 1971. *Physics and Beyond*. London: Allen and Unwin.
Jeans, Sir James, 1973. *The Mysterious Universe*. London: Pelican Books.
Kant, I., 1786. *Metaphysische Anfangsgründe der Naturwissenschaft*.
Langevin, P., 1950. *La pensée et l'action*. Paris: Ed. Français Réunis.
Lee, A. R., and T. M. Kalotas, 1975. *Am. J. Phys*. 43:434.
Leonardo da Vinci, *Scritti*. Milan: Instituto Editoriale Italiano.
Lévy-Leblond, J.-M., 1976a. "What Is So Special about Relativity?" in *Group Theoretical Methods in Physics*. Ed. A. Janner et al. Berlin: Springer Verlag.
—, 1976b. *Am. J. Phys*. 44:271.
—, 1977a. "Towards a Proper Quantum Theory," in *Quantum Mechanics, A Half-Century Later*, ed. J. Leite Lopes and M. Paty. Dordrecht: Reidel.
—, 1977b. "The Picture of the Quantum World: From Duality to Unity," *Int. J. Theor. Chem*. 12 (Suppl. 1).
—, 1979. "La relativité aujourd'hui," *La Recherche* 96/10.
—, 1980. "Speeds," *Am. J. Phys*. 48.
—, 1982. "In Praise of Wrong Theories," *Speculations in Science and Technology* 5.
Lévy-Leblond, J.-M., and J.-P. Provost, 1979. *Am. J. Phys*. 47.
Maitte, B., 1984. "Construction et applications de la notion de groupe en cristallographie." Lille: Université de Lille.
Poincaré, H., 1913. *La valeur de la science*. Geneva: Le Cheval Ailé.
Ravetz, J. R., 1972. "Galileo and the Mathematisation of Speed," in *La mathématisation des doctrines informes*, ed. C. Canguilhem. Paris: Hermann.
Shive, J. N., and R. L. Weber, 1982. *Similarities in Physics*. Bristol: A. Hilger.
Vogel, T., 1956. *Physique mathématique classique*. Paris: Armand Colin.
Weissman, Y., 1983. *Am. J. Phys*. 51.
—, 1985. *Am. J. Phys*. 53:17.
Wightman, A. S., 1976. "Mathematical Idealization and Theoretical Physics; the Role of Mathematical Physics," *Progress in Scientific Culture* 1:16.
Wigner, E. P., 1967. *Symmetries and Reflections*. Bloomington: Indiana University Press.
Yourgrau, W., and S. Mandelstam, 1968. *Variational Principles in Dynamics and Quantum Theory*. London: Pitman.

Why Does Physics Need Mathematics? A Comment

ITAMAR PITOWSKY

What is the source of "the unreasonable effectiveness of mathematics in physics"? Lévy-Leblond does not address this problem directly but notes in passing that "there is no need ... to marvel at the mysterious harmony and correspondence between mathematical structures and physical phenomena." The reason is that many, perhaps even a majority of pure mathematical structures, do not find their way into physical theories. Indeed, the ratio of mathematics to mathematical physics increases with time. In the central part of this comment I shall argue that on the contrary, there are good grounds to marvel at the aforementioned harmony.

Lévy-Leblond's observation, though correct, does not quite settle the issue. There are two reasons for this:

(a) It is impossible to foretell which portion of mathematics is relevant to physics and which is not. There is no clear a priori distinction between pure mathematics and mathematical physics.

(b) There is far more abstract mathematics involved in physics than one usually assumes.

Rather than arguing these points in the abstract, I shall present some not too well known examples, which are interesting in their own right.

At the beginning of the nineteenth century, Bolzano constructed an "unintuitive" example of a continuous function which is nowhere differentiable (see Kline 1972, chap. 40). Bolzano's work was ignored for some decades and was subsequently resurrected and expanded by Weierstrass around the middle of the nineteenth century. It should be stressed that before

E. Ullmann-Margalit (ed.) The Scientific Enterprise, 163–167.

that period, i.e., before the work of Cauchy and Weierstrass on the foundations of the infinitesimal calculus was published, there was no widely accepted rigorous definition of continuity. Instead, mathematicians relied on their physical and geometrical intuitions. This led to some amusing misunderstandings. Thus, the purpose of constructing continuous nowhere differentiable functions was to prove the need for a sharp distinction between continuity and differentiability, a distinction which was often blurred. The epistemic effect of this and similar work, if any, was in its demonstration that mathematics, as an art of pure forms, should follow its own rules, quite independently of physical intuition. This idea can also be evaluated in the broader context of the German Enlightenment, with its tendency to return to the Greek ideal of pure knowledge as opposed to the rather "deplorable" forms of practical applications.

Some fifty years after Weierstrass, in the miraculous year 1905, Einstein composed his paper on Brownian motion (Einstein 1926). Before this event, however, a few other physicists (Gouy and, in particular, F. M. Exner, around 1900)[1] attempted to explain this phenomenon on the basis of the same physical hypothesis, namely, that the motion of the macroscopic Brownian particle is caused by its collisions with the microscopic molecules of the medium. In their calculations, they relied on the observed average velocity of the Brownian particle: the net result was that their evaluation of Avogadro's number ($\sim 6.02 \times 10^{23}$) was off mark by five orders of magnitude, an extremely large error. Einstein, however, bypassed the reference to the particle velocity and obtained very accurate predictions.

At this point in time, N. Wiener entered the game with his mathematical theory of Brownian motion. Einstein, referring to Minkowski's work on special relativity, had once remarked ironically that "since the mathematicians have attacked the relativity theory I myself no longer understand it" (quoted in Clark 1972). In a sense, Wiener did for Brownian motion what Minkowski had done for relativity (though Wiener's work is bigger in scope and deeper).

Einstein's argument, noted Wiener, is really a probabilistic one. The probability space in question is the set of all possible (continuous) trajectories of the Brownian particle; the events, the presence of the particle in certain regions of space at certain times. On this so-called Wiener space there is a probability measure, the Wiener measure. One of the central theorems of this theory, due to Wiener and Paley, states the following: "With probability one, the path taken by a Brownian particle is nowhere differentiable" (formally: the set of all continuous paths which have a derivative even at a single point has Wiener measure zero).

As we see, the Brownian particle simply *does not have* a velocity. Thus, those "pathological" creatures, "continuous nowhere differentiable curves," were instrumental in explaining why a very large error had occurred in previous calculations. Needless to say, this analysis is only an idealization, which breaks down when one considers segments of the path which are of the order of magnitude of molecular dimensions. Both Einstein and Wiener were well aware of this. It turns out, however, that these considerations do not alter the quantitative aspects of the theory.

Nowhere differentiable functions and similar creatures play a role in more recent theories as well.[2] What originated as an argument against the primacy of physical intuition in mathematics has thus turned up in a physical theory. This is by no means an exception; non-Euclidean geometry is the best-known example, and there are many others. These examples indicate, I believe, that there exists no a priori principle which distinguishes pure mathematics from applied mathematics, mathematical physics in particular. Even the continuum hypothesis may make its appearance one day in a physical theory; I admit that this is a remote possibility, but I cannot conceive of a good a priori argument why it should not occur.

The fact that abstruse mathematics plays a subtle role in physical theories is not sufficiently appreciated. Take, for example, the concept of "real number," which has a direct bearing on our theories of the space-time continuum. There are many nonisomorphic models of the set of real numbers, some of which are quite bizzare from a topological or measure-theoretical perspective (see, e.g., Jech 1973). Thus, if one assumes that the denumerable axiom of choice is not valid, one can construct models of set theory in which the set of reals can be represented as a countable union of countable sets. In this model, Lebesgue measure is additive but not countably additive, a fact which undermines the validity of many important limit theorems. There are similar models in which the sequential and topological definitions of continuity do not coincide.

Most physicists are not even aware that such models exist, and for a good reason — these models are simply inadequate. But in their very (implicit) decision to disregard such "pathological" cases these physicists take for granted the validity of the (denumerable) axiom of choice. At least tacitly, then, this abstract set-theoretical consideration comes to play a role in physics.

I agree with Lévy-Leblond that one should not speak of the *application* of mathematics in physics, since physical thinking involves mathematics in an irreducible manner. But for this very reason, *because* of the entanglement of mathematical structure with physical fact, one should consider not only the relevant mathematical results but also the set of assumptions (axioms) underlying their proof. As a result, one discovers that "mathematical physics" is much broader than what is "officially" covered by that name.

All this raises questions as to the "correspondence" or "harmony" between mathematics and physics, and I think an explanation is called for. One possible route is to adopt a logicistic approach: mathematics is the study of all logical possibilities. The physical world imposes constraints on our free choice of merely consistent mathematical structures. Such a view also explains what Lévy-Leblond calls the "mathematical polymorphism" of physics. Physical constraints are never complete (in the technical sense), so that mathematically nonisomorphic but physically equivalent theories are always possible.

The trouble with this view lies in its diachronic nature. We are never presented with the class of all logical possibilities, nor can we ever be "presented" with such a class. The question is, therefore, what determines our ability to conceive of *logical* possibilities in the first place? Why is it that the above-mentioned "pathological" models of the reals were discovered (or invented?) only in the twentieth century, while "normal" models, with their implicit assumptions, have been available for hundreds of years? It seems that the physical world and physical intuition are partially responsible for our very ability to conceive of logical possibilities, and not only the reverse. I think that one cannot avoid the problem of "correspondence": one can understand it in a naive, realistic sense, or defend a thesis with a Kantian flavor, but ultimately one cannot escape metaphysics.

Lévy-Leblond uses the peculiar entanglement of mathematical structure and physical theory as a demarcation criterion between physics and the other natural sciences:

> It is through the role that mathematics plays in a given area of the natural sciences, according to its constitutive (or otherwise) character, that this area is recognized (or not) as belonging to the domain of physics.

I find this thesis very attractive. The following remarks are by way of clarification. First, the inevitable question: What makes a given use of mathematics constitutive, and which use is a mere (non-constitutive) application? I share the intuition that the use of mathematics in physics is "different" and deeper, but I find it difficult to provide an analytical frame for this observation.

The second (perhaps also inevitable) question is: How does this thesis fit in with the reductionist ethos of science? One of the central ideals in the scientific enterprise is to try and provide a unified explanation of all natural phenomena. The drawing of sharp lines of demarcation between the sciences appears to obstruct this idea, which has had some remarkable manifestations.

Notes

1. Exner 1900, 843. For more details see Nelson 1976.
2. For example, in the theory of fractals; see Mandelbrot 1977.

References

Clark, R. W., 1972. *Einstein, The Life and Times*. New York: Avon Books.

Einstein, A., 1926. *Investigations on the Theory of Brownian Motion*, ed. R. Furth. New York: Dover.

Exner, F. M., 1900. *Annalen der Physik* 2.

Jech, T. J., 1973. *The Axiom of Choice*. Amsterdam: North Holland.

Kline, M., 1972. *History of Mathematics from Ancient to Modern Times*. Oxford: Oxford University Press.

Mandelbrot, B., 1977. *Fractals, Form, Chance and Dimension*. San Francisco: W.H. Freeman.

Nelson, E., 1976. *Dynamical Theories of Brownian Motion*. Princeton, N.J.: Princeton University Press.

Analysis and Its Paradoxes

FELICIA ACKERMAN

Consider the following proposition.

(1) To be an instance of knowledge is to be an instance of justified true belief not essentially grounded in any falsehood.

If true, (1) illustrates an important type of philosophical analysis. For convenience of exposition, I will assume that (1) is a *correct* analysis. But nothing in my views about analysis hinges upon this. To suppose that (1) gives an analysis, however, leads to the notorious problem that philosophers nowadays are apt to call "the" paradox of analysis, as follows. If the property of justified true belief not essentially grounded in any falsehood is the *analysans* of the property of knowledge, it would seem that they are the same property, and hence that

(2) To be an instance of knowledge is to be an instance of knowledge

would have to be the same proposition as (1). But propositions (1) and (2) seem clearly diverse; for example, (2) is trivial (in the sense that understanding it entails believing it), while (1) can be informative and in fact was only recently arrived at by philosophers.

Classical writings on analysis suggest another paradox of analysis (see, e.g., Moore 1963b). Consider this proposition.

(3) An analysis of the property of being a brother is that to be a brother is to be a male sibling.

If (3) is true, it would seem that the property of being a brother would have to

E. Ullmann-Margalit (ed.) The Scientific Enterprise, 169–178.

be the same property as the property of being a male sibling, and that

> (4) An analysis of the property of being a brother is that to be a brother is to be a brother

would also have to be true — and in fact would have to be the same proposition as (3). Yet (3) is true and (4) is false.

Both these paradoxes rest upon the assumptions that analysis is a relation between concepts or properties, rather than one involving entities of other sorts, such as linguistic expressions, and that in a true analysis, *analysans* and *analysandum* will be the same property. (I am using the terms 'concept' and 'property' interchangeably in this paper.) Both these assumptions are explicit in Moore (1963b). But some of Moore's remarks hint at a solution — that a statement of an analysis is a statement partly about the concept involved and partly about the verbal expressions used to express it (Moore 1963b, 283–84). He says he thinks a solution of this sort is bound to be right, but fails to suggest one because he cannot see just how the analysis can be even partly about the expressions.

Elsewhere I have suggested a way and argued that this solves the second paradox for analyses that do not also raise the first paradox (Ackerman 1990). But this sort of solution could not possibly work for the first paradox.[1] The two paradoxes must be clearly distinguished. This can be done by means of the Frege-inspired sense-individuation condition, which is the condition that two expressions have the same sense if and only if they can be interchanged *salva veritate* whenever used in propositional attitude contexts. If 'knowledge' and 'justified true belief not essentially grounded in any falsehood' met this condition, (1) and (2) would not raise the first paradox; but the second paradox obviously can arise regardless of whether the senses of 'brother' and 'male sibling' are identical by this condition. The second paradox is a matter of the failure of such expressions to be interchangeable *salva veritate* in sentences purporting to give analyses. Thus, a solution aimed only at such sentences can solve the second paradox. This is emphatically false for the first paradox, however, which has analogues for all pairs of propostions in which *analysans* and *analysandum* are interchanged. It seems obviously unacceptable to hold that the proposition that John knows that his father is bald is partly about language because it can be believed without believing the proposition that John has justified true belief not essentially grounded in any falsehood that his father is bald.

The rest of this paper will deal only with the first paradox. Elsewhere, I have challenged Moore's assumption that *analysans* and *analysandum* must be the same concept and I have proposed a solution relying on the view that, in this sort of analysis, *analysans* and *analysandum* are diverse concepts but are related in a special epistemic way that I spell out (see Ackerman 1981; 1986). I say "this sort of analysis," because the term 'analysis' has obviously been used in philosophy in more than one sense. For example, what I have to say here has no bearing on reformatory analyses, whose aim is to replace a concept or expression considered in some way defective with one that is considered in the relevant way improved and hence is intended to be conceptually different from it. The paradox exemplified by (1) and (2) clearly does not apply to analyses of this sort, and from here on in this paper I am concerned only with the sort of analysis illustrated by (1). All my uses of the term 'analysis' in the following should be understood as restricted in this way.

The rest of the paper will proceed as follows. First I will outline the solution that I proposed elsewhere to the first paradox, and I will show how this solution leads to a new view of the nature of analysis, as well as having more general metaphysical and conceptual implications. Then I will extend my previous work, by drawing out some further epistemic aspects of the solution and discussing a problem that calls for its modification; this problem sheds new light on precisely what is supposed to be the relation between analysis and meaning on a view that denies conceptual identity between *analysans* and *analysandum*.

My solution accepts Moore's view that *analysans* and *analysandum* are concepts rather than linguistic expressions.[2] I have indicated that I reject Moore's view that *analysans* and *analysandum* are the same concept. The example of (1) makes it easy to see why such a rejection is intuitively satisfying, for the concepts of knowledge and justified true belief not essentially grounded in any falsehood seem on the face of it to be quite different. The first is readily understood, the second fairly difficult; the first is employed in the thoughts of many people; the second in those of relatively few, perhaps only professional philosophers and their students. But this type of solution also raises problems. In particular, it is unclear how two diverse concepts can be closely enough related to count as *analysans* and *analysandum*. Just what is this relation supposed to be? On the solution I proposed, the relation has the following facets.

(a) The *analysans* and *analysandum* are necessarily coextensive.
(b) The *analysans* and *analysandum* are knowable a priori to be coextensive.
(c) The *analysandum* is simpler than the *analysans* (a condition whose necessity is recognized in classical writings on analysis; see, e.g. Langford 1952). The notion of simplicity could doubtless profit

from further explication, but here I will leave it on the intuitive level.

(d) The *analysans* does not have the *analysandum* as a constituent.

Condition (d) rules out circularity. But since many valuable quasi-analyses are partly circular (e.g., knowledge is justified true belief supported by *known* reasons not essentially involving any falsehood), it seems best to distinguish between full analysis, for which (d) is a necessary condition, and partial analysis, for which it is not.

These conditions, while necessary, are clearly insufficient. The basic problem is that they apply to many pairs of concepts that do not seem closely enough related epistemically to count as *analysans* and *analysandum*, such as the concept '6' and the concept '$\sqrt[4]{1296}$'. Accordingly, my solution finds the fifth condition by drawing upon what actually seems epistemically distinctive about analyses of the sort under consideration — a certain way in which they can be justified. This is by the philosophical example and counterexample method, which in general terms goes as follows. J investigates the analysis of K's concept of Q (where K can but need not be identical to J) by setting K a series of armchair thought experiments, presenting K with a series of simple described hypothetical test cases, and asking K questions of the form: "If such-and-such were the case, would this count as a case of Q?" J then contrasts the descriptions of the cases where K answers affirmatively with the descriptions of the cases where K does not, and J generalizes upon these descriptions to arrive at the properties (if possible, not including the *analysandum*) and their mode of combination that constitute the *analysans* of K's concept of Q. Since J need not be identical with K, there is no requirement that K himself be able to perform this generalization, to recognize its result as correct, or even to understand the *analysans* which is its result. This is reminiscent of the observation that one can simply recognize a bird as a swallow without ever realizing just what features of the bird (beak, wing configuration, etc.) form the basis of this recognition.[3] K answers the questions on the sole basis of whether the described hypothetical cases just strike him as having Q. J observes certain strictures in formulating the cases and questions. He makes the cases as simple as possible, to minimize the possibility of confusion, and also to minimize the likelihood that K will draw upon his philosophical theories (or quasi-philosophical, rudimentary notions, if he is unsophisticated philosophically) in answering the questions. For this reason, if two hypothetical test cases yield conflicting results, the conflict should *ceteris paribus* be resolved in favor of the simpler case. J makes the series of described cases wide-ranging and varied, with the aim of its being a complete series; here I say that a series is complete if and only if no case that is omitted is such that, if included, it would change the analysis arrived at. J does not, of course, use as a test-case description anything complicated and general enough to express the *analysans*. There is no requirement that the described hypothetical test cases be formulated only in terms of what can be observed. Moreover, using described hypothetical situations as test cases enables J to frame the questions in such a way as to rule out, to a degree, extraneous background assumptions. Thus, even if K correctly believes that all and only P's are R's, the question of whether P, R, or both enter the *analysans* of his concept of Q can be investigated by asking him such questions as "Suppose (even if it seems preposterous to you) that you were to find out that there was a P that was not an R. Would you still consider it a Q?"

Taking all this into account, I suggest that the fifth condition for the *analysans-analysandum* relation is as follows.

> (e) If P is the *analysans* of Q, the proposition that necessarily all and only what is P is Q can be justified by generalizing from intuitions about the correct answers to questions of the sort indicated about a varied and wide-ranging series of simple described hypothetical situations.

(Of course, by the term 'intuition' I mean not any special and mysterious faculty, but merely one's beliefs about whether certain described hypothetical situations "just seem to be" or "just strike him as" cases of Q.)

I am glossing over some problems and qualifications that I discuss elsewhere with this view (Ackerman 1981, 1986). But two new points should be made about the special epistemic nature of the *analysans-analysandum* relation. One has to do with its peculiarly open-ended epistemic quality. Since analyses are justified by the hypothetical case, philosophical example and counterexample method, the sole ground for believing an analysis is that a wide-ranging series of such cases support it and no counterexamples seem to be forthcoming. This may seem to give analysis a quasi- (or more than quasi) empirical justification along the lines of "there have been no counterexamples after extensive investigation in the past; hence (by induction) there are unlikely to be any in the future." But it would be a mistake to view the matter in this way. Statements about experience or the past form no part of the justification of an analysis. The justification is simply that there seems to be wide-ranging support and no counterexamples. Yet it is true that analysis has no greater justification than a kind of induction based on the support of simple hypothetical cases and the absence of counterexamples.

This induction should not be considered empirical, however, since it is induction about cases that can be purely hypothetical, and hence it does not rely on empirical investigation.[4] But this characteristic of analyses does seem to make them, in a sense, more open-ended than necessary and a priori statements that can be proven deductively. Deductive proofs give a reason for accepting a conclusion that goes beyond a mere accumulation of cases, although it is of course possible to discover that one has made an error in a deduction, just as it is to discover that one has overlooked a counterexample to an analysis.

My second epistemic point here will lead into the problem for my account and to the more general question that I wish to discuss about the relation between analysis and meaning. The present point is this. On my view of analysis, statements giving analyses are not analytic! They are synthetic a priori. This is clear on standard accounts of analyticity. For example, if we use Quine's notion of an analytic truth as "one that can be turned into a logical truth by putting synonyms for synonyms" (Quine 1961, 23), propositions such as (1) will clearly not fit, since in my view the terms 'knowledge' and 'justified true belief not essentially grounded in any falsehood' express different concepts or properties and hence will not be synonyms. If we adapt the notion of 'conceptual containment' to propositions like (1), we get the same result. In my view, the concept of justified true belief not essentially grounded in any falsehood is not "conceptually contained" in the concept of knowledge.

Langford has also given an account of analysis which admits a species of analysis that is synthetic a priori (Langford 1952, 335). I have critized Langford's account elsewhere (Ackerman 1981). But it is clear that Moore required analyses to be analytic (see, e.g., Moore 1963b; 1963a, 235). And, of course, Moore equated indefinability and unanalyzability (e.g., Moore 1963a, 243). From the present standpoint, however, this is an error. So, although Moore takes his "open question argument" to show that the concept of good is unanalyzable and indefinable, on my view he errs in saying that this argument shows that good is unanalyzable. One could as well argue that the concept of knowledge was unanalyzable into the concept of justified true belief not essentially grounded in any falsehood, on the grounds that "whoever will attentively consider with himself what is actually before his mind when he asks the question: 'Is [justified true belief not essentially grounded in any falsehood] after all [knowledge]?' can easily satisfy himself that he is not merely wondering whether [knowledge] is [knowledge]."[5]

In my view, a term is indefinable if and only if it has no analysis that expresses the same property, and I say that indefinable terms express what I call irreducible properties. Hence, indefinability does not entail unanalyzability. For example, 'knowledge' is indefinable, but not unanalyzable. Thus, my view has implications that go beyond analysis and its paradoxes, since it holds that there are many irreducible concepts or properties — those whose analyses raise the apparent paradox of (1) — that philosophers have failed to take into account.

Good is unanalyzable in Moore's view; is it analyzable in mine? The issue raises a problem for my view, as follows. Suppose that J, like most people, does not believe any proposition of the form "All and only F things are intrisically good," where F is a nonnormative complex concept; but suppose that the fact that J counts all and only F things as intrinsically good can be discovered by the simple hypothetical test-case method of condition (e). If it is in fact necessary and knowable a priori that all and only F things are intrinsically good, then the concept F seems to count, on my foregoing account, as the *analysans* of J's concept of intrinsically good. (Whenever I use the term 'good' henceforth in this paper, it should be understood in the sense of 'intrinsically good.') But now suppose that a person K bears the same relation to the proposition that all and only G things are good that J bears to the proposition that all and only F things are good, where G is also a nonnormative complex concept and the concepts F and G are not even extensionally equivalent, let alone identical. A question arises: do J and K have the same concept of good? It seems unreasonable to say they do not, simply on the grounds that they count different things as good, since they may still be able to argue with each other about moral issues, understand each other's moral discourse, etc. But if they have the same concept of good, what, if anything, is its analysis? If only J is considered, my foregoing account would yield the view that the concept F is the *analysans,* but once K is investigated, this seems wrong.

Of course, one possible move here is to deny that sentences of the form "All and only F things are good," where F is a nonnormative concept, actually express propositions that are necessary and can be known a priori to be true. But it is undesirable to rest a view of analysis on such a controversial assumption. So I will consider two possible approaches to the issue. Although neither approach will preserve *analysans-analysandum* concept identity, the approaches differ as to just how close they hold the relation between analysis and meaning to be.

The first approach is to change analysis from a two-place relation between an *analysans* and an *analysandum* to a three-place relation among an *analysans*, an *analysandum*, and a person. Thus, one concept could have different analyses for different people. On this view, property P will be the *analysans* of Q for J if and only if conditions (a)–(e), as modified in the following way, are satisfied. Conditions (c) and (d) are unchanged; (e) is restricted specifically to generalizations from J's intuitions about correct answers to questions about the appropriate sort of described hypothetical situations; and conditions (a) and (b) are replaced with the proposition that J treats the *analysans-analysandum* equivalence as a priori in the sense that the generalizations from his intuitions about the simple described hypothetical cases support it even when the cases are set up so as to rule out empirical background assumptions in the way illustrated above. This, of course, does not require that J has the belief that the *analysans-analysandum* equivalence is a priori, but it allows the *analysans* for J of the concept of good to be what is expressed by a description that tells us what sorts of things J counts as good, regardless of whether it is true that all and only such things are good.

This approach may leave some residual feelings of dissatisfaction. I think they are connected with the idea that such descriptions as "most conducive to the general happiness" cannot express the *analysans* of the concept of good for anyone, precisely because such descriptions are not closely enough connected to what 'good' means. J's understanding a term does not entail his knowing its analysis, but it may seem that the sense in which it entails his ability to use the term correctly is that, if the term is analyzable, its *analysans* can be arrived at by generalizing from his intuitions about the correct answers to the sort of questions mentioned in condition (e). This seems to be supported by standard cases of analysis. For example, we would be inclined to say that someone whose replies to the appropriate sorts of questions supported the generalization that knowledge is true belief, whether justified or not, did not quite have our concept of knowledge. The fact that people with different intrinsic moral beliefs do not automatically have different concepts of good might be connected with the issue of analysis by returning to the original ideas of analysis as a two-place relation between *analysans* and *analysandum*, but adding the following to my original list of necessary and sufficient conditions for an analysis.

(f) If P is the *analysans* of Q and generalizations from J's intuitions about the correct answers to questions of the sort mentioned in (e) do not support the view that necessarily all and only cases of P are cases of Q, then J does not count as using the word in question to express the concept Q.

(He may, of course, be using it to express a closely related concept. For simplicity of exposition, I am assuming here, of course, that J is using all the other relevant expressions in the standard way.)

Note that, since my view of analysis does not claim that analysis is a matter of meaning or concept identity, there is of course no circularity in relying on these notions to explicate my view. Note, also, that this second approach would seem to hold that, although Moore was correct in asserting the unanalyzability of good, he was correct for the wrong reason. With this approach, we get what is likely to be a general impossibility of analyzing normative concepts in nonnormative terms. This can be illustrated even with the standard example of knowledge, whose analysis requires the normative concept of justification. This raises another issue about the sorts of cases referred to in condition (e). Just as people with incompatible intrinsic moral beliefs can have the same concept of good, an epistomological skeptic and an antiskeptic can have the same concept of knowledge, since skeptical arguments derive whatever initial plausibility they may have precisely by drawing upon our ordinary concepts. But suppose that J is a skeptic, not just in a reflective philosophical way, but even in his basic intuitions about particular cases. Then his answer to such questions as, "if you stuck your head out of the window and saw and felt rain, would you know it was raining?" will be negative. But this would be because, for him, adequate justification would supposedly be lacking. It would not mean that for him the analysis of knowledge was different from the usual one. This case shows the importance of including normative concepts in the test-case descriptions of the sort mentioned in condition (e) when the *analysandum* is itself normative.

Notes

1. I am not suggesting that Moore would expect it to. His whole discussion of analysis seems more oriented toward the second paradox. (See Moore 1963b.)
2. For simplicity of exposition in this paper, I will sometimes speak of the analysis of expressions, but I should always be understood as considering these expressions as expressing certain properties — the properties they actually do express. With this proviso, we can say much the same things about the analysis of expressions and the analysis of concepts.
3. The philosophical implications of this way of recognizing swallows are discussed by Walton (1972).
4. There is an interesting parallel here with the fallacious argument that mathematical proofs are really empirically justified because in performing them we rely on pens and paper, perceptions, etc. One way to expose the fallacy there is to note that it is in no way essential to the proofs that the paper and pens really exist rather than, for example, are imagined.
5. Moore 1963a, 242. I have replaced Moore's references in this passage to "good" and "pleasure" with references to "knowledge" and "justified true belief not essentially grounded in any falsehood."

References

Ackerman, F., 1981. "The Informativeness of Philosophical Analysis," *Midwest Studies in Philosophy* 6:313–20.

—, 1986. "Essential Properties and Philosophical Analysis," *Midwest Studies in Philosophy* 2:305–13.

—, 1990. "Analysis, Language and Concepts: The Second Paradox of Analysis," in *Philosophical Perspectives,* ed. J. Tomberlin, vol. 4. Ataskadero, Calif.

Langford, C. H., 1952. "The Notion of Analysis in Moore's Philosophy," in *The Philosophy of G. E. Moore*, ed. P. Schilpp. New York.

Moore, G. E., 1963a. "The Indefinability of Good," in *Readings in Twentieth Century Philosophy*, ed. W. Alston and G. Nakhnikian. London.

—, 1963b. "A Reply to My Critics: Analysis," in *Readings in Twentieth Century Philosophy,* ed. W. Alston and G. Nakhnikian, London, 279–85.

Quine, M. V., 1961. "Two Dogmas of Empiricism," in idem, *From a Logical Point of View*. Cambridge, Mass.: Harvard University Press, 20–46.

Walton, K., 1972. "Linguistic Relativity," in *Conceptual Change*, ed. G. Pearce and P. Maynard. Dordrecht: D. Reidel Publishing Company, 1–29.

Explanation and Practical Reason

CHARLES TAYLOR

I

Our modern conceptions of practical reason are shaped — I might say are distorted — by the weight of moral skepticism. Even conceptions which intend to give no ground to skepticism have frequently taken form in order to resist it best, or to offer the least possible purchase to it. In this, practical reason falls into line with a pervasive feature of modern intellectual culture, which one could call the primacy of the epistemological: the tendency to think out the question "what something *is*" in terms of the question "how it is *known*."

The place of what I call skepticism in our culture is evident. By this I do not mean just a disbelief in morality, or a global challenge to its claims — though the seriousness with which a thinker like Nietzsche is regarded shows that this is no marginal position. I am also thinking of the widespread belief that moral positions cannot be argued; that moral differences cannot be arbitrated by reason; that when it comes to moral values, ultimately we all just have to plump for the ones which feel/seem best to us. This is the climate of thought which Alasdair MacIntyre calls (perhaps a bit harshly) "emotivist" (MacIntyre 1981, ch. 2), which at least ought to be called in some sense "subjectivist." If you ask any undergraduate class of beginners in philosophy, the majority will claim to adhere to some form of subjectivism. This may not correspond to deeply felt convictions. It does seem to reflect, however, what these students imagine the intellectually respectable option to be.

E. Ullmann-Margalit (ed.) The Scientific Enterprise, 179–201.

What underpins this climate? Some rather deep metaphysical assumptions, when one gets down to it. But certainly, on the immediate level, it is fostered by the actual experience of moral diversity. On an issue like abortion, for instance, it does not seem to be possible for either side to convince the other. Protagonists of each tend to think that their position is grounded on something self-evident. For some, it just seems clear that the fetus is not a person, and it is absurd to ruin the life of some being who undeniably has this status in order to preserve it. For others, it is absolutely clear that the fetus is both alive and human, and so destroying it cannot be right, unless murder is. Neither side would budge from these initial intuitions, and once one accepts either one, the corresponding moral injunctions seem to follow.

If the seemingly helplessness of reason tells us something about its real limits, then a worrying thought arises: what if some people came along who failed to share our most basic and crucial moral intuitions? Suppose some people thought that innocent human beings should be killed in order to achieve some advantage for the others, or make the world more aesthetically pleasing, or something of the sort? And haven't we actually experienced people who stepped way outside the bounds of our core morality: the Nazis, for instance? Is reason as powerless before such people as it seems to be to arbitrate the dispute about abortion? Is there no way to prove them wrong?

Here is where our implicit model of practical reason begins to play an important role. If "showing them" means presenting facts or principles which they cannot but accept, and which are sufficient to disprove their position, then we are indeed incapable of doing so. But one could argue that that is a totally wrong view of practical reason. Faced with an opponent who is *unconfusedly* and *undividedly* convinced of his position, one can indeed only hope to move him rationally by arguing from the ground up, digging down to the basic premises we differ on, and showing him to be wrong there. But is this really our predicament? Do we really face people who quite lucidly reject the very principle of the inviolability of human life?

In fact, this does not seem to be the case. Intellectual positions put forward to justify behavior like that of the Nazis — to the extent that any of their ravings justify this appellation at all — never attack frontally the ban on murder of conspecific. They are always full of special pleadings: e.g., that their targets are not really of the same species, or that they have committed truly terrible crimes which call for retaliation, or that they represent a mortal danger to others, etc. This kind of stuff is usually so absurd and irrational that it comes closer to raving than to reason. And, of course, with people on this kind of a trip, reason is in fact ineffective as a defense. But this is not to say that reason is powerless to show them wrong. Quite the contrary. The fact that these terrible negations of civilized morality depend so much on special pleading, and of a particularly mad and irrational sort, rather suggests that there are limits beyond which *rational* challenges to morality have great trouble reaching.

This might indicate a different predicament of, and hence task for, practical reasoning. Its predicament would be defined by the fact that there are limits to what people can unconfusedly and undividedly espouse; so that, for instance, in order to embrace large-scale murder and mayhem, they have to talk themselves into some special plea of the sort mentioned above, which purports to square their policies with some recognized version of the prohibition against killing. But these pleas are vulnerable to reason, and in fact barely stand up to the cold light of untroubled thought.

The task of reasoning, then, is not to disprove some radically opposed first premise (e.g., killing people is no problem), but rather to show how the policy is unconscionable on premises which both sides accept, and cannot but accept. In this case, its job is to reveal the special pleas.

On this model — to offer here at any rate a first approximation — practical argument starts off on the basis that my opponent already shares at least some of the fundamental dispositions toward good and right which guide me. The error comes from confusion, unclarity, or an unwillingness to face some of what he cannot lucidly repudiate; and reasoning aims to show up this error. Changing someone's moral view by reasoning is always at the same time increasing his self-clarity and self-understanding.

Here are two quite different models of practical reason; let us call them the apodeictic and the ad hominem models, respectively. John Stuart Mill was making use of a distinction of this kind, and opting for the second, in his famous (perhaps notorious) remarks in *Utilitarianism*. "Questions of ultimate ends are not amenable to direct proof," he avers, and yet "considerations may be presented capable of determining the intellect either to give or to withhold its assent to the doctrine [sc. of utility]; and this is the equivalent to proof" (Mill 1979, 4–5). This may sound like someone trying to squirm his way out of a contradiction, but the distinction is quite clear and sound. One cannot argue someone into accepting an ultimate end if he really rejects it. However, the whole case of utilitarians is, in fact, that people *do not* reject it, that they all really operate by it, albeit in a confused and therefore self-defeating fashion. And this is why there may be "considerations ... capable of determining the intellect." Mill shows us, in fact, what he thinks these are in chapter IV, where he goes on to argue that what people in fact desire is happiness (Mill 1979, 34). The appeal is to what the opponent already seeks, a clear view of which has to be rescued from the confusions of intuitionism.

But, it might be thought, this invocation of Mill is enough to discredit the ad hominem model irremediably. Is this not exactly where Mill commits the notorious "naturalistic fallacy," arguing from the fact that men desire happiness to its desirability, on a glaringly false analogy with the inference from the fact that men see an object to its visibility (Mill 1979, 34)? Derisive hoots echo through philosophy classes since Moore, as beginning students cut their teeth on this textbook example of a primitive logical error.

There is no doubt that this argument is not convincing as it stands. But the mistake is not quite so simple as Moore claimed. The central point indicated by the Moorean objection is the special nature of moral goals. This is a phenomenon which I have tried to describe with the term "strong evaluation" (cf., e.g., Taylor 1985). Something is a moral goal of ours not just by virtue of the fact that we are de facto committed to it. It must have this stronger status, that we see it as demanding, requiring, or calling for this commitment. While some goals would have no more claim on us if we ceased desiring them — e.g., my present aim to have a strawberry ice cream cone after lunch — a strongly evaluated goal is one such that, were we to cease desiring it, we would be shown up as insensitive, brutish, or morally perverse.

That is the root of our dissatisfaction with Mill's argument. We feel that just showing that we always desire something, even that we cannot help desiring it, does nothing by itself to show that we *ought* to desire it, that it is a moral goal. Suppose I were irremediably addicted to smoking; would that prove that I ought to smoke? Clearly not. We always understand smoking as a weakly evaluated end. We have to distinguish between showing with regard to some end that we cannot help desiring it, and showing that all our strong evaluations presuppose it, or involve it, once we overcome our confusions about them. In the second case, we would have demonstrated that we cannot be lucid about ourselves without acknowledging that we value this end. This is the sense in which it is inescapable, not after the fashion of some de facto addiction. Whereas addictions are rightly declared irrelevant to moral argument, except perhaps negatively, the proof of inescapable commitment is of the very essence of the ad hominem mode of practical reasoning, and is central to the whole enterprise of moral clarification.

Mill is plainly onto some intuition to this effect in deploying the argument in *Utilitarianism*. One of the things he is trying to show is that everyone else's commitments collapse into his. But the argument is botched up because of a crucial weakness of the doctrine of utility itself, which is based on the muddled and self-defeating attempt to do away with the whole distinction between strong and weak evaluation. The incoherence of Mill's defense of the "higher" pleasures on the grounds of mere de facto preference by the "only competent judges" (Mill 1979, 8–11) is also a testimony to the muddles and contradictions to which this basically confused theory gives rise.

But this does point to one of the most important roots of modern skepticism. We can already see that people will tend to despair of practical reason to the extent that they identify its mode of argument as apodeictic. This clearly sets for it an impossible task. But this model will be accepted to the degree that the alternative, the ad hominem one, appears inadequate or irrelevant. And this it is bound to do, as long as the distinction between strong and weak evaluation is muddled over or lost from sight. This confusion can only breed bad arguments à la Bentham and Mill, and these, once denounced, discredit the whole enterprise.

But utilitarianism does not come from nowhere. The whole naturalist bent of modern intellectual culture tends to discredit the idea of strong evaluation. The model for all explanation and understanding is the natural science which emerges out of the seventeenth-century revolution. But this offers us a neutral universe; it has no place for intrinsic worth, or for goals which make a claim on us. Utilitarianism was partly motivated by the aspiration to build an ethic which was compatible with this scientific vision. But to the extent that this outlook has a hold on the modern imagination, even beyond the ranks of utilitarianism, it militates in favor of accepting the apodeictic model, and hence of a quasi-despairing acquiescence in subjectivism.

The link between naturalism and subjectivism is even clearer from another angle. The seventeenth-century scientific revolution destroyed the Platonic-Aristotelian conception of the universe as the instantiation of Forms, which defined the standards by which things were to be judged. The only plausible alternative construal of such standards in naturalist thought was as projections of subjects. They were not part of the fabric of things, but rather reflected the way subjects react to things, the pro/con attitudes that they adopt. Now perhaps it is a fact that people's attitudes tend to coincide — a happy fact, if true; but this does nothing to show that this point of coincidence is more correct than any other possible one.[1]

The opposition to this naturalist reduction has come from a philosophical stance which might in a broad sense be called "phenomenological." By this I mean a focus on our actual practices of moral deliberation, debate, and understanding. The attempt is to show, in one way or another, that the vocabularies we need to explain human thought, action and feeling, or to explicate, analyze, and justify ourselves or each other, or to deliberate on what to do, all rely inescapably on strong evaluation. Or, put negatively, that the attempt to separate out a language of neutral description which combined with commitments or pro/con attitudes might recapture and make sense of our actual explanations, analyses, deliberations, etc., must always lead to failure. It seems to me that this case has been convincingly made in a host of places.[2]

This kind of argument is, of course, not only a justification of the very foundation of the ad hominem model of reasoning, but an example of it. It tries to show us that in all lucidity we cannot understand ourselves, or each other, nor can we make sense of our lives or determine what to do, without accepting a richer ontology than naturalism allows and thinking in terms of strong evaluation. This might be thought to beg the question, establishing the validity of a mode of argument through a use of it. But the presumption behind this objection ought to be challenged: what ought in fact to trump the ontology implicit in our best attempts to understand/explain ourselves? Should the epistemology derived from natural science be allowed to do so, so that its metaphysical bias in favor of a neutral universe overrules our most lucid self-understandings in strongly evaluative terms? But does this not rather beg the crucial question, viz., whether and to what extent human life is to be explained in terms modeled on natural science? And what better way to answer this question than by seeing what explanations actually wash?

II

Enough has been said above to show that one of the strongest roots of modern skepticism/subjectivism in regard to ethics is the naturalist temper of modern thought. This tends to discredit in advance the ad hominem mode of argument, which actually might hold out the hope of settling certain moral issues by reason, and leaves in the field only the apodeictic model, which clearly sets an impossible standard. Within a human situation inescapably characterized in strongly evaluative terms, we can see how argument aimed at self-clarification might, in principle at least, bring agreement. In a neutral universe, whatever agreement there is between attitudes seems merely a brute fact, irrelevant to morals, and disagreement seems utterly inarbitrable by reason, bridgeable only by propaganda, arm-twisting, or emotional manipulation.

But this analysis brings to mind another source of modern skepticism, constituted by the independent attractions of the apodeictic model itself. Here is where we really measure the tremendous hold of epistemology over modern culture.

This model emerges *pari passu* with and in response to the rise of modern physical science. As we see it coming to be in Descartes, and then Locke, it is a foundationalist model. Our knowledge claims are to be checked, to be assessed as fully and responsibly as they can be, by breaking them down and identifying their ultimate foundations, as distinct from the chain of inferences which build from these toward our original unreflecting beliefs. This foundationalist model can easily come to be identified with reason itself. Modern reason no longer tends to be understood substantively but procedurally, and the procedures of foundationalism can easily be portrayed as central to it. But from the foundationalist perspective, only the apodeictic mode of reasoning is really satisfactory; the appeal to shared fundamental commitment seems simply a recourse to common prejudices. The very Enlightenment notion of prejudice encapsulates this negative judgement.

This brings us to another aspect. Foundationalist reasoning is meant to shake us loose from our parochial perspective. In the context of seventeenth-century natural science, this involved, in particular, detaching us from the peculiar human perspective on things. The condemnation of secondary qualities is the most striking example of this move to describe reality no longer in anthropocentric but in "absolute" terms.[3]

But if the canonical model of reasoning involves maximally a breaking us free from our perspective, then the ad hominem model cannot but appear inferior, since by definition it starts from what the interlocutor is already committed to. And here a particularly important consideration comes into play. Starting from where your interlocutor is not only seems an inferior mode of reason in general, but it can be presented as a peculiarly bad and, indeed, vicious form of practical reason. For all those whose instinct tells them that the true demands of morality require radical change in the way things are and in the way people have been trained to react to them, starting from the interlocutor's standpoint seems to be a formula for conservatism, for stifling at the onset all radical criticism, and for foreclosing all the really important ethical issues.

This has always been one of the strongest appeals of utilitarianism, and one of the greatest sources of self-congratulation by partisans of utility. It is not only that their theory has seemed to them the only one consonant with science and reason, but also that they alone permit reform. Mill argues against views based on mere "intuition" that they freeze our axiomata media forever, as it were, and make it impossible to revise them, as mankind progresses and our lights increase. "The corollaries from the principle of utility,... admit of indefinite improvement, and, in a progressive state of the human mind, their improvement is perpetually going on" (Mill 1979, 24).

Here is a source of modern skepticism and subjectivism which is as powerful as naturalism, and tends to operate closely in tandem with it, viz., the belief that a critical morality, by its very nature, rules out the ad hominem mode of practical reasoning. Naturalism and the critical temper together tend to force us to recognize the apodeictic mode as the only game in town. The obvious severe limitations of this mode in face of ethical disagreement then push us toward a half-despairing, half-complacent embracing of an equivocal ethical subjectivism.

I have tried to show elsewhere that this identification of the demands of critical morality with a procedural understanding of reason and the apodeictic mode is deeply mistaken.[4] But erroneous or not, it clearly has been immensely influential in our intellectual culture. One can see this in the way people unreflectingly argue in terms of this model.

Discuss the question of arbitrating moral disputes with any class, graduate or undergraduate, and very soon someone will ask for "criteria." What is aimed at by this term is a set of considerations such that, for two explicitly defined, rival positions *X* and *Y*, (a) people who unconfusedly and undividedly espouse both *X* and *Y* have to acknowledge them, and (b) they are sufficient to show that *Y* is right and *X* is wrong, or vice versa. It is then driven home, against those who take an upbeat view of practical reason, that for any important moral dispute, no considerations have both (a) and (b). If the rift is deep enough, things which are (b) must fail of (a), and vice versa.

The problem lies with the whole unreflecting assumption that "criteria" in this sense are what the argument needs. As we explore this further, we shall see that this assumption, as it is usually understood in the context of foundationalism, amounts to ruling out the most important and fruitful forms of the ad hominem mode.

The whole assumption that rational arbitration of differences needs "criteria" has become very problematic, and not only for practical reasons. It is also a notorious source of puzzlement and skeptical challenges in the history of science. It is some underlying assumption of this kind which has driven so many people to draw skeptical conclusions from the brilliant work of Thomas Kuhn (conclusions to which Kuhn himself has sometimes been drawn, without ever succumbing to them). For what Kuhn persuasively argued was the "incommensurability" of different scientific outlooks which have succeeded each other in history. That is, their concepts are non-intertranslatable and, what is even more unsettling, they differ as to what features or considerations provide the test of their truth. The considerations each recognizes as having (b) are diverse. There are no criteria. And so the radical inference of Feyerabend has seemed widely plausible: "anything goes."

But as Alasdair MacIntyre has argued in another work (MacIntyre 1977), it is clear that what needs revision here is our metatheory of scientific reasoning, rather than, e.g., our firmly established conviction that Galileo made an important step forward relative to Aristotelian physics. The blind acceptance of a foundationalist, apodeictic model of reasoning is perhaps just as damaging here as in ethics. Recalling how inadequate the model is can both help to weaken its hold on us in general, and allow us to see more exactly what is truly peculiar to practical reason.

MacIntyre argues very convincingly that the superiority of one scientific conception over another can be rationally demonstrated, even in the absence of what are normally understood as criteria. These are usually seen as providing some externally defined standard, against which each theory is to be weighed independently. But what may be decisive is that we be able to show that the *passage* from one to the other represents a gain in understanding. In other words, we can give a convincing narrative account of the passage from the first to the second as an advance in knowledge, a step from a less good to a better understanding of the phenomena in question. This establishes an asymmetric relation between them: a similarly plausible narrative of a possible transition from the second to the first could not be constructed. Or, put in terms of a real historical transition, portraying it as a loss in understanding is not acceptable.[5]

What I want to take from this is the notion that one can sometimes arbitrate between positions by portraying *transitions* as gains or losses, even where what we normally understand as decision through criteria — qua externally defined standards — is impossible. I should like to sketch here three argument forms, in ascending order of radical departure from the canonical, foundationalist mode.

1. The first takes advantage of the fact that we are concerned with transitions, that the issue here is a comparative judgment. On the standard, unreflecting assumptions of foundationalism, comparative judgments are usually secondary to absolute ones. Rival positions *X* and *Y* are checked against the facts, and one is shown superior to the other because it predicts or explains certain facts which the other does not. The comparative judgment between the two is based on absolute judgments concerning their respective performance in face of reality. The role of criteria here is taken by facts, observations, protocols, or perhaps by standards to be applied to explanations of facts, such as elegance or simplicity. Just as in a football game, the comparative verdict — team *X* won — is founded on two absolute assessments; e.g., team *X* scored 3 goals and team *Y* scored 2 goals. The most popular theory of scientific reasoning with this traditional structure, that of Popper, resembles indeed the eliminative rounds in a championship. Each theory plays the facts, until it suffers defeat and is then relegated.

But, as MacIntyre shows, comparative reasoning can draw on more resources than this. What may convince us that a given transition from *X* to *Y* is a gain is not only or even so much how *X* and *Y* deal with the facts, but how they deal with each other. It may be that, from the standpoint of *Y*, not just the phenomena in dispute, but also the history of *X* and its particular pattern of anomalies, difficulties, makeshifts, and breakdowns can be greatly illuminated. In adopting *Y*, we make better sense not just of the world, but of our history of trying to explain the world, part of which has been played out in terms of *X*.

The striking example, which MacIntyre alludes to, is the move from Renaissance sub-Aristotelian to Galilean theories of motion. The Aristotelian conception of motion, which entrenched the principle which claims that there is no motion without a mover, ran into tremendous difficulty in accounting for "violent" motion, e.g., the motion of a projectile after it leaves the hand or cannon mouth. The Paduan philosophers, and others, looked in vain for factors which could play the continuing role of movers in pushing the projectile forward. What we now see as the solution was not forthcoming until theories based on inertia altered the entire presumption of what needs explaining: continued rectilinear (or, for Galileo, circular) motion is not an *explanandum*.

What convinces us still today that Galileo was right can perhaps be put in terms of the higher "score" of inertial theories over Aristotelian ones in dealing with the phenomena of motion. After all this time, the successes of the former are only too evident. But what also was, and still is, an important factor — and which obviously bulked relatively larger at the time — is the ability of inertial theories to make sense of the whole pattern of difficulties which beset the Aristotelians. The superiority is registered here not simply in terms of their respective "scores" in playing "the facts," but also by the ability of each to make sense of itself and the other explaining these facts. Something more emerges in their stories about each other than was evident in a mere comparison of their several performances. This shows an asymmetric relation between them: you can move from Aristotle to Galileo, realizing a gain in understanding, but not vice versa.

2. This is still not a radical departure from the foundational model. True, decisive criteria are not drawn from the realm of facts, or of universally accepted principles of explanation. But, the crucial considerations are still accessible to both sides. Thus, the pre-Galileans were not unaware of the fact that they had a puzzling problem with violent motion. To use Kuhnian terms, this was an "anomaly" for them, as testified by their intellectual perplexity and the desperate expedients they resorted to. The decisive arguments are transitional; they concern what each theory has to say about the other, and about the passage from its rival to itself, which takes us beyond the traditional way of conceiving validation, both positivist and Popperian. But, in the strict sense of our definition above, there are still "criteria" here, for the decisive considerations are such that both sides must recognize their validity.

All the same, it can be argued that if we look at the seventeenth-century revolution from a broader perspective this ceases to be so. Thus, if we stand back and compare the dominant models of science before and after the break, we can see that different demands were made on explanation. The notion of a science of nature, as it came down from Plato, and especially from Aristotle, made the explanation in terms of Forms (*eide* or species) central, and beyond that posited an order of forms, whose structure could be understood teleologically, in terms of some notion of the good or of what ought to be. Such principles as plenitude, which Lovejoy identifies and traces, make sense on the understanding that we can know beforehand, as it were, that the universe will be so ordered as to realize the maximum richness (Lovejoy 1960). Similarly, explanations in terms of correspondences are possible, since it follows from the basic conception that the same constellation of ideas will be manifested in all the different domains.

Now, if science consists of a grasp of order of this kind, then the activity of explaining why things are as they are (what we think of as science) is intrinsically linked to the activity of determining what the good is, and in particular how human beings should live through attuning themselves to this order. The notion that explanation can be distinct from practical reason, that the attempt to grasp what the world is like can be made independent of the determination of how we should stand in it, that the goal of understanding the cosmos can be uncoupled from our attunement to it, makes no sense in the premodern understanding.

But, notoriously, the seventeenth-century revolution brought about an uncoupling of just this kind. The turn to mechanism offers a view of the universe as neutral; within it, cause-effect relations can be exploited to serve more than one purpose. We might say that Galileo and his successors turn toward an utterly different paradigm of explanation. If scientific explanation can always be roughly understood as rendering in some sense the puzzling comprehensible by showing how the phenomenon to be explained flows from mechanisms or modes of operation which we understand, then the seventeenth century sees a massive shift in the kind of understanding which serves as the basic reference point.

There is certainly one readily available mode of human understanding which the Platonic-Aristotelian tradition drew on. We are all capable of understanding things in terms of their place in a meaningful order. These are the terms by which we explain events as the initially puzzling behavior of others, or social practices which seemed strange at first, or some of the odd-seeming details of a new work of art and so on. In another different sense of "understanding," we understand an environment when we can find our way in it, get things done in it and effect our purposes in it. This is the kind of understanding a garage mechanic has, and I unfortunately lack, of the environment under the hood of my car.

One way of describing the seventeenth-century revolution in science is to say that a paradigm of understanding comes to take the place of another as the basic reference point for scientific explanation of nature.[6] But this has as an ineluctable consequence the redemption of the explanation from practical reason, mentioned above. Only the first type of understanding lends itself to a marriage of the two.

But, once we describe it in this way, the scientific revolution can be made to appear as not fully rationally motivated. Today we all accept that Galileo was right. But can we *justify* that preference in reason? Was the earlier outlook shown to be inferior, or did its protagonists just die out? If you ask the ordinary person today for a quick statement as to why he holds modern science as superior to the premodern, he will probably point to the truly spectacular technological payoff that has accrued to Galilean science. But here is where the skeptic can enter. Technological payoff, or the greater ability to predict and manipulate things, is certainly a good criterion of scientific success on the post-Galilean paradigm of understanding. If understanding is knowing your way about, then modern technological success is a sure sign of progress in knowledge. But how is this meant to convince a pre-Galilean? For, in fact, he is operating with a quite different paradigm of understanding, to which manipulative capacity is irrelevant, and which rather proves itself through a different ability, that of discovering our proper place in the cosmos and finding attunement with it. And, it could be argued, modern technological civilization is a spectacular failure at *this*, as ecological critics and green parties never tire of reminding us.

Is the argument then to be considered a standoff between the two, judged at the bar of reason? Here the skeptical spin-off from Kuhn's work makes itself felt. Once one overcomes anachronism and comes to appreciate how different earlier theories were and how great the breaks are in the history of knowledge — and this has been one of the great contributions of Kuhn's work — then it can appear that no *rational* justification of the transitions is possible. Each theory carries with it its own built-in criteria of success — moral vision and attunement in one case, manipulative power in the other — and is therefore invulnerable to the other's attack. In our period, the choice seems to be manipulative power, but this has to be for some extra-epistemic consideration, i.e., not because this mode of science has been shown superior as *knowledge*. Presumably, we prefer that payoff. In terms of my earlier discussion, what we lack here are "criteria"; there are no decisive considerations which *both* sides must accept.

Some people are driven by their epistemological position to accept some account of this kind.[7] But this seems to me preposterous. Once more, it can appear plausible only because it fails to think of the transition between the two views. It sees each as assessing a theory's performance in face of reality by its own canons. It does not go further and demand of each that it give an account of the existence of the other; i.e., not just explain the world, but explain also how this rival (and presumably erroneous) way of explaining the world could arise.

Once one makes this demand, one can appreciate the weakness of pre-Galilean science. There is a mode of understanding which consists of knowing one's way about. This is universally recognized. In making another mode paradigm for scientific explanation, pre-Galilean science drew on a set of assumptions which entailed that this manipulative understanding would never have a very big place in human life. It always allowed for a lower form of inquiry, the domain of "empirics," who scramble around to discover how to achieve certain effects. But the very nature of material embodiment of Forms, as varying, approximate, and never integral, ensured that no important discoveries could be made here, and certainly not an exact and universal body of findings. Consequently, the very existence of such a body of truths, and the consequent spectacular manipulative success, represents a critical challenge to premodern science. Indeed, it is difficult to see how it could meet this challenge. On its basic assumptions, modern science should not have got off the empiric's bench, emerged from the dark and smelly alchemist's study to the steel-and-glass research institutes that design our lives.

So the problem is not some explanatory failure on its own terms, not some nagging, continuing anomaly, as in the narrower issue (above) of theories of motion; it is not that pre-Galilean science did not perform well enough by its own standards; or that it does not have grounds within itself to downgrade the standards of its rivals. If we imagine the debate between the two theories as being carried on endlessly on Olympus, without any actual results emerging, then it is indeed a stand-off. But what the earlier science cannot explain is the very success of the later science *on the later's own terms*. Beyond a certain point, you just can no longer pretend that manipulation and control are not relevant criteria of scientific success. Pre-Galilean science died of its inability to explain/assimilate the actual success of post-Galilean science, where there was no corresponding symmetrical problem. And this death was quite rationally motivated. On Olympus the grounds would have been insufficient; but faced with the actual transition, you are ultimately forced to read it as a gain. Once again, what looks like a stand-off when two independent closed theories are confronted with the facts, turns out to be conclusively arbitrable by reason when you consider the transtion.[8]

I have been arguing above that the canonical, foundationalist notion of arbitrating disputes through criteria generates skepticism about reason, which disappears once one sees that one is often arguing about transitions. And we have seen that this skepticism affects some of the more important transitions of science just as much as it does the disputes of morality — and for the same reason, viz., the seeming lack of common criteria. In particular, it tends to make the history of science seem less rational than it has in fact been.

The second case is a more radical departure from the canonical model than the first. For the defeat does not come from any self-recognized anomaly in the vanquished theory. Nevertheless, there was *something* which the losing theory had to recognize outside the scope of its original standards, viz., that the very success of mechanistic science posed a problem. If we ask why this is so, we are led to recognize a human constant, viz., a mode of understanding of a given domain, *D*, which consists of our ability to make our way about and effect our purposes in *D*. We might borrow a term from Heidegger and call this understanding, as we originally have it prior to explicitation or scientific discovery, "pre-understanding." One of the directions of increasing knowledge of which we are capable consists of making this pre-understanding explicit, and then extending our grasp of the connections which underlie our ability to deal with the world as we do. Knowledge of this kind is intrinsically linked with increased ability to effect our purposes, with the acquisition of potential recipes for more effective practice. In some cases, it is virtually impossible to extend such knowledge without making new recipes available; and an extension of our practical capacities is therefore a reliable criterion of increasing knowledge.

Because of these links between understanding and practical ability, we cannot deny whatever increases our capacities the title of being a gain in knowledge in some sense. We can seek to belittle its significance, or deem it to be by nature limited, disjointed, and lacunary, as Plato does. But then we have to sit up and take notice when it manages to burst the bounds we set for it; and this is what has rendered the transition to Galilean science a rational one.

The mediating element is something deeply embedded in the human life-form, of which we are all implicitly aware, and which we have to recognize when made explicit: the link between understanding (of a certain kind) and practical capacity. But then is not the predicament of reason here coming to look analogous to the description offered above of moral disputes? The task is not to convince someone who is undividedly and unconfusedly attached to one first principle that he ought to shift to an entirely different one. So described, it is impossible. Rather, we are always trying to show that, granted what our interlocutor already accepts, he cannot but attribute to the acts or policies in dispute the significance we are urging on him.

Now here it has been a question of altering the first principles of science — the paradigms of understanding underlying it and the standards of success. And we can see a rational path from one to the other, but only because in virtue of what the pre-Galilean already accepts, he cannot but recognize the significance of the Galilean science's massive leap forward. In both cases it is not a question of radical conversion from one ultimate premise to the other. That would indeed be irrational. Rather we show that the pre-Galilean could not undividedly and unconfusedly repudiate the deliverances of post-Galilean science as irrelevant to the issue that divides them.

Perhaps, then, those ultimate breakpoints we refer to as "scientific revolutions" share some logical features with moral disputes. They are both rendered irrational and seemingly inarbitrable by an influential but erroneous model of foundationalist reasoning. To understand what reason can do in both contexts, we have to see the argument as being about transitions. And, as the second case makes plain, we have to see it as making appeal to our implicit understanding of our form of life.

This brings to the fore one of the preconceptions which has bedeviled our understanding here and fostered skepticism. On the standard foundationalist view, the protagonists are seen as closed explicit systems. Once one has articulated their major premises, it is assumed that all possible routes of appeal to them have been defined. So the pre-Galilean model, with its fixed standards of success, is seen as impervious to the new standards of prediction and control. But the real positions held in history do not correspond to these watertight deductive systems, and that is why rational transitions are in fact possible.

We could argue that there are also moral transitions which could be defended in an analogous way to the scientific one just described. When one reads the opening pages of Foucault's *Surveiller et Punir* (Paris 1975), with its riveting description of the torture and execution of a parricide in the mid-eighteenth century, one is struck by the cultural change we have gone through in post-Enlightenment Western civilization. We are much more concerned about pain and suffering than our forebears. We shrink more from the infliction of gratuitous suffering. It would be hard to imagine people taking their children to such a spectacle in a modern Western society, at least openly and without some sense of unease and shame.

What has changed? It is not that we have embraced an entirely new principle, or that our ancestors would have thought the level of pain irrelevant and providing no reason at all to desist from some course of action involving torture or wounds. It is, rather, that this negative significance of pain was subordinated to other, weightier considerations. If it is important that punishment in a sense undoes the evil of the crime and restores the balance — which is implicit in the whole notion of the criminal making *amende honorable* — then the very horror of parricide calls for a particularly gruesome punishment. It calls for a kind of a theater of the horrible as the medium in which the undoing can take place. In this context, pain takes on a different significance: there has to be a lot of it to make it work. The principle of minimizing pain is trumped.

It is then possible to see how the transition might be assessed rationally. If the whole outlook which justifies trumping the principle of minimizing suffering — which involves seeing the cosmos as a meaningful order in which human society is embedded as a microcosm or mirror — comes to be set aside, then it is rational to be concerned above all with reducing suffering. Of course, our ultimate judgment will depend on whether we see the change in cosmology as rational; and that is, of course, the issue that I have been arguing above in connection with the scientific revolution. If I am right there, then here too the transition can perhaps be justified.

I am not claiming that the only factor involved in this important change has been the decline of the earlier cosmology. There are other, independent grounds in modern culture which have made us more reluctant to inflict pain. Some of them may have sinister aspects, if we believe Foucault himself. It is not possible to discuss all this here,[9] but it is essential to recognize the decline of the older notion of cosmic/social order as *one* consideration which lends a rational grounding to modern humanitarianism. This change would not only be linked to that in scientific theory; it would also be analogous to it in rational structure: to something which has always been recognized, although formerly in a subordinate place (the link between understanding and practice, the good of reducing pain), we are now constrained to give a more central significance because of changes which have taken place.

The analogy I have been drawing between the justification of some scientific and moral revolutions cannot hide the fact that a great many moral disputes are much more difficult to arbitrate. To the extent that one can call on human constants, these are much more difficult to establish. The suspicion dawns that in many cases such constants are of no avail. The differences between some cultures may be too great to make any ad hominem form of argument valid between them. Disputes of this kind would be inarbitrable.

3. This form of argument, from the constants implicitly accepted by the interlocutor, does not exhaust the repertoire of practical reason. There is one more form, which is also an argument about transitions, but is an even more striking departure from the canonical model. In both the above forms the winner has appealed to some consideration which the loser had to acknowledge — his own anomalies, or some implicit constant. In the light of this consideration, it was possible to show that the transition from *X* to *Y* could be shown as a gain, but not the reverse. So there is still something like a criterion operating here.

We can imagine a form of argument in which no such consideration is invoked. The transition from *X* to *Y* is not shown to be a gain because this is the only way to make sense of the key consideration; rather, it is shown to be a gain directly, because it can plausibly be described as mediated by some error-reducing move. This third mode of argument can be said to reverse the direction of argument. The canonical foundationalist form can only show that the transition from *X* to *Y* is a gain in knowledge by showing that say, *X* is false and *Y* true, or *X* has probability *n*, and *Y* has 2*n*. The two forms we have been considering focus on the transition, but they too only show that the move from *X* to *Y* is a gain, because we can make sense of this transition from *Y*'s perspective but not of the reverse move from *X*'s perspective. We still ground our ultimate judgment in the differential performance of *X* and *Y*.

Let us consider the possibility that we might identify the transition directly as the overcoming of an error. If we knew that it consisted in the removing of a contradiction, or the overcoming of a confusion, or the recognition of a hitherto ignored relevant factor, the order of justifying argument would be reversed. Instead of concluding that *Y* was a gain over *X* because of the superior performance of *Y*, we would be confident of the superior performance of *Y* because we knew that *Y* was a gain over *X*.

Are we ever in a position to argue in this direction? In fact, examples abound in everyday life. First take a simple case of perception. I walk into a room, and see, or seem to see, something very surprising. I pause, shake my head, rub my eyes, and place myself to observe carefully. Yes, there is really a pink elephant with yellow polka dots in the class. I guess someone must be playing a practical joke.

What has gone on here? In fact, I am confident that my second perception is more trustworthy, not because it scores better than the first on some measure of likelihood. On the contrary, if what I got from the first look was something like: "maybe a pink elephant, maybe not," and from the second, "definitely a pink elephant with yellow polka dots," there is no doubt that the first must be given greater antecedent probability. It is after all a disjunction, one of whose arms is overwhelmingly likely in these circumstances. In fact I trust my second percept, because I have gone through an ameliorating transition. This is something I know how to operate, it is part of my know-how as a perceiver. That is what I in fact create by shaking my head (to clear the dreams), rubbing my eyes (to get the rheum out of them), and setting myself to observe with attention. It is my direct sense of the transition as an error-reducing one which grounds my confidence that my perceptual performance will improve.

Something similar exists in more serious biographical transitions. Joe was previously uncertain whether he loved Anne or not, because he also resented her, and in a confused way he was assuming that love is incompatible with resentment. But now he sees that these two are distinct and compatible emotions, and the latter is no longer getting in the way of his recognizing the strength of the former. Joe is confident that his present self-reading ("I certainly love Anne") is superior to his former self-reading ("I'm not sure whether I love Anne"), because he knows that he has passed from one to the other via the clarification of a confusion, i.e., a move which in its very nature is error-reducing.

Some of our gains in moral insight prove themselves to us in just this way. Pete was behaving impossibly at home, screaming at his parents, acting arrogantly with his younger siblings, and he felt resentment all the time and was very unhappy. He felt a constant sense of being cheated out of his rights. That at least is how it was formulated by his parents to the social worker. Now things are much better. Pete now applies the above description himself to his former feelings. In a confused way, he felt that something more was owed to him as the eldest, and he resented not getting it. But he would never have subscribed to any such principle, and he clearly wants to repudiate it now. He thinks his previous behavior was unjustified, and that one should not behave that way toward people. In other words, he has gone through a moral change; his views of what people owe each other in the family have altered. He is confident that this change represents moral growth, because it came about through dissipating a confused, largely unconsciously held belief, one which could not survive his recognizing its real nature.

These three cases are all examples of my third form of argument. They are all biographical. They deal with transitions of a single subject, whereas the standard disputes I have been discussing fall between people. They are often (in the first case, always) cases of inarticulate, intuitive confidence; and hence, arguably, they have nothing to do with practical *reason* at all, if this is understood as a matter of forms of *argument*.

These two points are well taken. I have chosen the biographical context because this is where this order of justification occurs at its clearest. The same form can be, and is, adapted to the situation of interpersonal argument. Imagine I am a parent, or the social worker, reasoning with Pete before the change. Or say I am a friend of Joe's talking out his confused and painful feelings about Anne. In either case, I shall be trying to offer them an interpretation of themselves which identifies these confused feelings as confused, and which thus, if accepted, will bring about the self-justifying transition.

This is, I believe, the commonest form of practical reasoning in our lives, where we propose to our interlocutors transitions mediated by such error-reducing moves, by the identification of contradiction, the dissipation of confusion, or by rescuing from (usually motivated) neglect a consideration whose significance they cannot contest. But this is a form of argument where the appeal to "criteria," or even to the differential performance of the rival views in relation to some decisive consideration, is quite beside the point. The transition is justified by the very nature of the move which effects it. Here the ad hominem mode of argument is at its most intense, and most fruitful.

III

In conclusion, I would like to try to draw together the threads of this perhaps too rambling discussion. I argued at the outset that practical arguments are in an important sense ad hominem. As a first approximation, I described these as arguments which appeal to what the opponent is already committed to, or at the least cannot lucidly repudiate. The notion that we might have to convince someone of an ultimate value premise which he undividedly and unconfusedly rejects is indeed a ground for despair. Such radical gaps may exist, particularly between people from very different cultures. In such cases, practical reason is certainly powerless.

But the discussion in the second part allows us to extend our notion of this kind of argument. It is not just cases where we can explicitly identify the common premise from the outset that allow of rational debate. This was in fact the case with my opening example. Both a Nazi and myself accept some version of the principle "Thou shalt not kill," together with a different set of exclusions. Rational argument can turn on why he can permit himself the exclusions he does. In fact, this historic position does not stand up very long to rational scrutiny. It was really mob hysteria masquerading as thought.

But our discussion of transitions shows how debate can be rationally conducted even where there is no such explicit common ground at the outset.

These arguments, to the effect that some transition from *X* to *Y* is a gain, are also ad hominem, in two related ways. First, they are specifically directed to the holders of *X,* in a way that apodeictic arguments never are. A foundational argument to the effect that *Y* is the correct thesis shows its superiority over the incompatible thesis *X* only incidentally. The proof also shows *Y*'s superiority over all rivals. It establishes an absolute, not just a comparative, claim. If I establish that the correct value for the law of attraction is the inverse square and not the inverse cube of the distance, this also rules out the simple inverse, the inverse of the fourth power, etc.

It is crucial to transition arguments that they make a more modest claim. They are inherently comparative. The claim is not that *Y* is correct *simpliciter,* but simply that, whatever is "ultimately" true, *Y* is better than *X*. It is, one might say, less false. The argument is thus specifically addressed to the holders of *X*. Its message is that whatever else turns out to be true, you can improve your epistemic position by moving from *X* to *Y*; this step is a gain. Nothing need follow from this for the holders of third, independent positions. Above all, there is no claim to the effect that *Y* is the ultimate resting point of inquiry. The transition claim here is perfection compatible with a further one which might one day be established, identifying a new position *Z*, which in turn supersedes *Y*. As MacIntyre puts it,

> We are never in a position to claim that now we possess the truth or now we are fully rational. The most that we can claim is that this is the best account which anyone has been able to give so far, and that our beliefs about what the marks of "a best account so far" are, will themselves change in what are at present unpredictable ways. (MacIntyre 1977, 455)

Second, these arguments all make their case by bringing to light something that the interlocutor cannot repudiate. Either they make better sense of his inner difficulties than he can himself (case 1), or they present him with a development which he cannot explain on his own terms (case 2), or they show the transition to *Y* to happen through a move which is intrinsically described as error-reducing (case 3). But in relation to the original example of arguing with a Nazi, these greatly extend the range of rational debate. For what they appeal to in the interlocutor's own commitments is not explicit at the outset, but has to be brought to light. The pattern of anomalies and contradiction only becomes clear, and stands out as such, from the new position (case 1); the full significance of a hitherto marginalized form of understanding only becomes evident when the new position develops it (case 2); that my present stance reposes on contradiction, confusion, and screening out the relevant only emerges as I make the transition — indeed, in this case, making the transition is just coming to recognize this error (case 3).

In other words, the range of rational argument is greatly extended once we see that not all disputes are between fully explicated positions. Here the canonical foundationalist model is likely to lead us astray. As we saw above with the second case, pre-Galilean science is indeed impregnable if we just think of its explicit standards of success: it has no cause to give any heed to technological payoff. This payoff, in fact, constitutes a devastating argument, to which one can only do justice by articulating implicit understandings which have hitherto been given only marginal importance. I would argue now that a great deal of moral argument involves the articulation of the implicit, and this extends the range of the ad hominem far beyond the easy cases where the opponent offers us purchase in one of his explicit premises.

I want to end with the basic claim with which I started, and which underlies this exploration. It is that modern philosophy, and to some extent modern culture, has lost its grip on the proper patterns of practical reason. Moral argument is understood according to inappropriate models, and this naturally leads to skepticism and despair, which in turn has an effect on our conception of morality, gives it a new shape (or misshapes it). We are now in a better position to see some of the motivations of this misunderstanding.

I believe that we can identify in the above discussion three orders of motivation which combine to blind us. First, the naturalist temper, with its hostility to the very notion of strong evaluation, tends to make the ad hominem argument seem irrelevant to ethical dispute. To show that your interlocutor is really committed to some good proves nothing about what he ought to do. To think it does is to commit the "naturalistic fallacy."

Second, naturalism together with the critical temper have tended to brand ad hominem arguments as illegitimate. Reason should be as disengaged as possible from our implicit commitments and understandings, as it is in natural science, and as it must be if we are not to be victims of the status quo with all its imperfections and injustices. But once we neutralize our implicit understandings, by far the most important field of moral argument becomes closed and opaque to us. We lose sight altogether of the articulating function of reason.

This distorts our picture not only of practical reason, but also of much scientific argument. This brings us to the third motive: the ascendancy of the foundationalist model of reasoning which comes to us from the epistemological tradition. This understands rational justification as (a) effected on the basis of "criteria," (b) judging between fully explicit positions, and (c) yielding in the first instance absolute judgments of adequacy/

inadequacy, and comparative assessments only mediately from these. We have just seen what an important role in our reasoning is played by irreducibly comparative judgments — judgments about transitions — by articulating the implicit, and by the direct characterization of transitional moves which make no appeal to criteria at all. To block all this from view through an apodeictic model of reasoning is to make most moral discussion incomprehensible. It also does not leave unimpaired our understanding of science and its history as we have amply seen. The connections are in fact close between scientific explanation and practical reason: to lose sight of one is to fall into confusion about the other.

Notes

1. See J. L. Mackie, *Ethics* (London: Penguin Books, 1977), for an excellent example of the consequences of uncompromising naturalist thinking.
2. Cf., e.g., McDowell, "Virtue and Reason," *The Monist* 62 (1979): 331–50, and also MacIntyre 1981. I have also tried to contribute to this in my "Neutrality in Political Science," in *Philosophical Papers*, Vol. 2 (Cambridge, 1985).
3. I borrow the term from the interesting discussion in Bernard Williams, *Descartes* (London: Penguin Books, 1978), 66–67. I have discussed this issue in my "Self-Interpreting Animals," in *Philosophical Papers*, Vol. 1 (Cambridge, 1985).
4. "Justice after Virtue," unpublished MS.
5. See a parallel notion of the asymmetrical possibilities of transition, this time applied to practical reason, in Ernst Tugendhat's notion of a possible "Erfahrungsweg" from one position to another, in *Selbstbewusstsein und Selbstbestimmung* (Frankfurt, 1979), 275.
6. I am borrowing here from Max Scheler's analysis in his essays: "Soziologie des Wissens" and "Erkenntnis und Arbeit," in *Werke*.
7. For instance, this seems to be implicit in Mary Hesse's view: see her "Theory and Value in the Social Sciences," in *Action and Interpretation*, C. Hookway and P. Petit, eds. (Cambridge, 1979), 2. She speaks there of prediction and control as "pragmatic" criteria of scientific success.
8. I have discussed this point at somewhat greater length in my "Rationality," in *Rationality and Relativism*, M. Hollis and S. Lukes, eds. (Cambridge, 1982); reprinted in my *Philosophical Papers*, vol. 2.
9. I have discussed this at greater length in my "Foucault on Freedom and Truth," *Political Theory* 12 (May 1984), 20.

References

Lovejoy, A., 1960. *The Great Chain of Being*. New York: Harper Torchbook.
MacIntyre, A., 1977. "Epistemological Crises, Grammatic Narrative, and the Philosophy of Science," *The Monist* 60: 453–72.

—, 1981. *After Virtue*. Notre Dame, Ind.: University of Notre Dame Press.
Mill, J. S., 1979. *Utilitarianism*. Indianapolis: Hackett.
Taylor, C., 1985. "What Is Human Agency?" in *Philosophical Papers*, Vol. 1. Cambridge: Cambridge University Press.

Symposium

"The Other Newton"

The Theological and Alchemical Writings

Gravity and Alchemy

B. J. T. DOBBS

Introduction

My earlier studies have yielded evidence that Newton was concerned from the first in his alchemical work in finding verification for the existence of a vegetative principle operating in the natural world, a principle he understood to be the secret, universal, animating spirit of which the alchemists spoke. He saw analogies between the vegetative principle and light and between the alchemical process and the work of the Deity at the time of creation. It was by use of this active vegetative principle that God constantly molded the universe to his providential design, producing all manner of generations, resurrections, fermentations, and vegetations. In short, it was the action of the secret animating spirit of alchemy that kept the universe from being the sort of closed mechanical system for which Descartes had argued (Dobbs 1975, 1979, 1982).

That Newton was committed to the alchemical enterprise is no longer denied. However, once that proposition is admitted, a broader question immediately springs to mind: What role, if any, did Newton's alchemical commitment play in the other areas of his work? In particular, I will here examine the role of alchemy in Newton's conceptualization of gravity as an active principle. Westfall has long argued for a connection between gravity and alchemy in this matter of active principles (Westfall 1971, and subsequent publications) and increasing familiarity with Newton's alchemical papers has substantiated his view.

E. Ullmann-Margalit (ed.) The Scientific Enterprise, 205–222.

However, the relationship between gravity and alchemy in Newton's thought was more complex than it once seemed, and this paper will examine that relationship in the 1660s and 1670s in some detail, concluding with a very brief sketch of events after 1680.

Gravity and Alchemy in the 1660s: Mechanical and Neoplatonic Influences

At first in the 1660s Newton considered gravity to be a mechanical mode of action. He began to ponder the problem about 1664. The editors of the undergraduate notebook that records Newton's early ideas suggest influences from Descartes, Digby, and Boyle on his first mechanical theory, which was derivative and nonmathematical.

> The matter causing gravity must pass through all the pores of a body. It must ascend again, (1) for either the bowels of the Earth must have large cavities and inanities to contain it, (2) or else the matter must swell the Earth.... For it must descend very fast and swift as appears by the falling of bodies and by the great pressure toward the Earth. It must ascend in another form than it descends, or else it would have a like force to bear bodies up as it has to press them down.... The stream descending will grow thicker as it comes nearer the Earth; but it will not lose its swiftness until it finds as much opposition as it has help from the flood following behind it. But when the streams meet on all sides in the midst of the Earth, they must needs be compressed into a narrow room, closely pressed together, and thus very much opposing one another either turn back the same way they came, or crowd through one another's streams with much difficulty and pressure, and so be compacted, by continually pressing them to the Earth until they arise to the place from whence they came. There they will attain their former liberty. (McGuire and Tamny, 1983, 363, 365, 427)

Though the precise form is Newton's own, his theory is a restatement of impact physics, conventional, orthodox mechanical philosophy — in short, an ether theory of gravitation. Bodies descend through the impulsion of fine material particles; a mechanical stream of ethereal matter causes gravity.

No hint exists in Newton's earliest statement of what gravity is to become for him later: an active principle directly dependent upon God's will. Yet even in the student notebook there is a recognition of God's omnipresence, the omnipresence that is later to subsume universal gravity. Violent motion in a plenum requires that some matter be crowded out of the way of a moving body, Newton said; therefore the motion meets with resistance. Not so with violent motion in a vacuum.

> But in vacuo it meets with nothing impenetrable to stay it. It is true God is as far as vacuum extends, but he, being a spirit and penetrating all matter, can be no obstacle to the motion of matter; no more than if nothing were in its way. (Ibid., 409.)

Later Newton will repeat his conviction that God is present where there is no body, as well as where body is also present. There, as here, God penetrates all matter (cf. Grant 1981, 227–28, on Henry More's belief in "All-penetrating" deity). But whereas later the omnipresence of God and his ability to penetrate matter have the utmost significance with respect to gravity, that is not the case in the student notebook, where gravity is caused by the mechanical motion of small particles of matter, to which God's presence simply constitutes "no obstacle."

Some five years later the first active principle appears in Newton's papers, in his alchemical "Propositions" of about 1669 (Keynes Collection, MS. 12A). In the "Propositions" Newton argues for the existence of a vital agent diffused through all things. The agent is "the mercurial spirit, most subtle and wholly volatile." Though it does have a "general method of operating in all things," it also has "particularities" of operation according to its subject matter, and it is responsible for both putrefaction and generation. The mercuriality, subtility, and volatility of the vital "spirit" seem to imply that it is material, though barely so. It has nothing to do with gravity, only with the phenomena of life and death.

Though Newton's synopsis in the "Propositions" came entirely from the alchemical tradition, as his own annotations in the manuscript indicate, his interest in the alchemical vital spirit was probably stimulated by Neoplatonic critiques of Descartes' mechanical system. Descartes had attributed too much efficacy to brute matter moving by ordinary mechanical laws. If the original common matter of the universe only moved "naturally," there could be no differentiation, neither sun, nor sky, nor earth, nor vortices, nor any heterogeneous thing, as Henry More had been quick to observe in a letter to Descartes himself.[1] More in his later publications, as well as Ralph Cudworth in his *True Intellectual System of the Universe* (1678), forced the issue of the inadequacy of a purely mechanical system. The realm of matter was the realm of necessity, of fate, of nature. To the Platonists matter had the lowest ontological standing and could not by itself perform the higher operations associated with providential design. For design, spiritual guidance of material particles was a necessity. A "spirit of nature" or a "soul of the world," intermediate between God and matter, handled such guidance and organized matter according to a divine plan (Cassirer 1953; Hutin 1966; McGuire 1968,

1977). Newton fully accepted the program of the Cambridge Platonists, such as More and Cudworth, not rejecting mechanical philosophy per se, but desiring to modify it to include spiritual principles capable of initiating motion in passive matter and then guiding it into providentially designed configurations. But, not content with a philosophical statement of that program, Newton turned to a study of alchemy in order to implement it, for alchemy not only discussed active principles but also claimed to provide experimental evidence of their operation.

McGuire has argued that the more general metaphysical tradition of Neoplatonism provides an adequate explanation of Newton's use of active principles in his later work and that there is thus no need to see alchemy as playing an essential role (McGuire 1977, 140–41, n. 84). From the strictly philosophical point of view, McGuire makes a legitimate point, but his argument is not tenable from the historical point of view. As a matter of historical record, Newton's papers demonstrate that he used alchemical experimentation to search for the facts regarding active principles. With those facts he hoped to found a more adequate natural philosophy, which included not only mechanical principles but spiritual ones as well.

In any event, there is one way in which Newton's earliest work on active principles differs from that of the Neoplatonic philosophers, a difference having to do with the materiality of the alchemical spirit. The most likely source of Newton's philosophical stimulus in the 1660s was Henry More's *Immortality of the Soul* (1659); the guiding "spirit of nature" More postulated there was decisively incorporeal (More 1659, 450; Burtt 1954, 135–43). But as we have seen, Newton's alchemical spirit had elements of materiality with its subtility, mercuriality, and volatility. He called it a "spirit," as he had called God a "spirit" in the student notebook, but seventeenth-century "spirits" were notoriously ambiguous, existing in a broad gray area between solid matter and deity. They could be either corporeal or incorporeal, as required by the exigencies of the problem to be explained (Pagel 1958, passim; McMullin 1978, 54–56; Emerton 1984, 179–93). But for Newton it seems that God was the only purely incorporeal spirit; all created spirits had some degree of corporeality. That had not been the case with More, for whom created spirits as well as the deity were incorporeal. Had Newton deliberately broken with More on this point by the end of the 1660s? It is more likely that Newton was only recording in the "Propositions" the characteristics of the vital spirit as they were reported in the alchemical literature. Although degrees of corporeality were later to be of great concern to him, there is no evidence that he attempted to resolve such issues in the 1660s.

Gravity and Alchemy in the 1670s: Stoic Influences

In the 1670s Newton began to weave the two concepts together — the concept of the material, mechanical ether that was the cause of gravity, and the spiritual, yet still material, agent of vitality. He seems to have made the relatively simple assumption that the more "spiritual" an agent is, the finer the matter of which it is composed, or, conversely, the smaller the material particle, the greater its "spirituality" and activity. That appears to be the assumption that allowed him to treat his mechanical and nonmechanical agents in a comparable format during the ensuing decade and even combine their functions in the economy of nature. For, as we shall see, Newton thought of gravity as mechanical during part of this period, while at the same time characterizing the vital spirit as nonmechanical or "vegetable."

Among the Stoics, an assumption similar to Newton's was made, relating decreasing materiality with increasing activity. In ascending from the earth through water and air to fire in the conventional ranking of the elements, the Stoics considered elementary matter to become more tenuous, but at the same time more tense and active. Even more significantly, in Stoic philosophy all elementary bodies continued to bear within themselves, in their innermost constitution, traces of the divine and almost totally immaterial creative fire that had formed the universe and that was the most tense and active substance of all. Distinguished from the elementary fire that can only burn and destroy (that is, fire conceived as an element or qualified body), the hidden creative fire of the Stoics (that is, fire conceived as the active principle) was the material agent for the shaping of matter, but it was also identified with the Stoic deity (Lapidge 1973; Hahm 1977; Hager 1982). The alchemists were often very close to that Stoic distinction between fiery principle and fiery element, insisting that their occult fire, their hidden active principle, was by no means to be equated with ordinary fire, for the latter could only burn and destroy while "their" fire was a creative one (Taylor 1953). Newton spoke of nature's secret fire in his "Vegetation" manuscript, describing it as the only principle of vegetation. And among his later alchemical notes, entitled "Out of La Lumière," we find just such a concept embodied in an alchemical description of the creation. Following the descending Stoic order for the creation of the elements from fiery creative principle to air, water, etc., Newton recorded the following scheme.

> The vapour of ye Elements wch is very pure & almost insensible & conteins in it the spirit of fire or light, wch is the form of the Univers & being so impregnated wth the spirit of ye universe represents the first Chaos conteining all things necessary to ye creation, that is the matter

> universal & form universal. And in descending & becoming sensible it first puts on ye body of ye air wch we breath & becomes inclosed in it to nourish & vivify all nature. And that it may act more easily upon the grosser bodies of vegetables & minerals it becomes still denser & insinuates it self into the water. (Yahuda MS. Var. 1, Newton MS. 30, f. 2v)

Newton's assumption that "spirituality" and activity are inversely proportional to materiality thus could have been rooted in alchemy, in Stoic philosophy, or in both. Whatever its ultimate source, it was an assumption that served him not only in the 1670s but beyond. Translated into the terminology of force, the same assumption appears in one of the Queries to the *Opticks* in Newton's public presentation of a late speculative ether for the explanation of gravity: "The exceeding smallness of its particles may contribute to the greatness of the force by which those particles may recede from one another . . ." (*Opticks,* 352), and McGuire has noted a number of closely related draft Queries to which the same idea is fundamental (Portsmouth MS. Add. 3970, quoted in McGuire 1977, 160–61).

In the early 1670s the mechanical gravitational ether and nonmechanical alchemical spirit appear partially fused in Newton's treatise "Of nature's obvious laws and processes in vegetation" (Burndy MS. 16). In this important manuscript the matter causing gravity is thoroughly "entangled" with active spirits, yet in this same set of papers, Newton also distinguishes with some care the natural processes he thinks are "mechanical" from those he thinks are "vegetable." That the problem of gravity is taken up at all in this primarily alchemical document is of supreme significance, and we must follow his thoughts on it in detail.

Newton was apparently led to a reconsideration of his early gravitational ether through a discussion of its "grosser" counterpart of atmospheric air. Air is generated by several processes, he said, the chief of which are "mineral dissolutions and fermentations" in the earth. Being then "protruded" by more air ascending under it, the air rises higher and higher until it "straggle into the aetherial regions," carrying with it vapors, exhalations, and even whole clouds (Burndy MS. 16, f. 3v). The air ascending in this speculative scenario is evidently that matter in a "grosser consistency" than ether, of which Newton spoke in his student notebook. It needs be grosser, so it will not press bodies upward with as great a force as the finer ether presses them down.

In the "Vegetation" manuscript Newton proceeded to a quantitative estimate of the air thus being continuously generated, then noted that when air reaches the ethereal regions it presses on the ether.

> This [the air] constantly crowding for room, the aether will become pressed thereby and so forced continually to descend into the earth from whence the air come; and there 'tis gradually condensed and interwoven with bodies it meets there and promotes their actions, being a tender ferment. (Ibid.)

In this brief passage there are three points of special interest. First, particles of air and ether change places with each other by "crowding." The motion is a circulatory one, not unlike that of the Cartesian vortices. Second, the limits for the rising air are set by the "aetherial regions," which are at no great distance from the earth. Thus the whole system seems to be earthbound, an interpretation reinforced by the recollection of the frequent references to the earth in the student description, where "the bowels of the Earth" must have many cavities or else (1) swell with the ethereal flood or (2) extrude the ethereal matter again in another form. Such an interpretation is further suggested by a related passage in the "Vegetation" manuscript, in which the earth "draws in ethereal breath... and transpires again with gross exhalations" (ibid.). Third, and perhaps most important for our understanding of Newton's evolving thought, is the fact that the "tender ferment," the vital and vegetable principle, has come to be associated with gravitational ether. The descending ether becomes "interwoven" with bodies and "promotes their actions," i.e., it initiates or supports activity in ordinarily passive matter.

Not that the gravitational function of the ether has been displaced. By no means, for Newton continued:

> But in its descent it [the aether] endeavors to bear along what bodies it passes through, that is, makes them heavy, and this action is promoted by the tenacious elastic constitution, whereby it takes the greater hold on things in its way, and by its vast swiftness. So much aether ought to descend as air and exhalations ascend, and therefore the aether being by many degrees more thin and rare (as air is than water), it must descend so much the swifter and consequently have so much more efficiency to drive bodies downward than air has to drive them up. (Ibid.)

Newton still thought at that time that his gravity was mechanical in its operation, for he explicitly observed later in the manuscript that "Nature's actions are either vegetable or purely mechanical (gravity, flux, meteors, vulgar chemistry)" (ibid., f. 5r). But in fact, though his ethereal particles do still act primarily by impact, they have also acquired a "tenacious elastic constitution" that enables them to take a "greater hold" on objects. Those new properties add an element to gravitational action that is not strictly mechanical, one that is soon to become more prominent in Newton's speculations, as we shall see.

But early in the 1670s, in the "Vegetation" manuscript, gravity is still primarily mechanical and it also still seems to be earthbound. Newton continued his discussion of it in a way that apparently referred it only to the earth.

> And this is very agreeable to nature's proceedings, to make a circulation of all things. Thus this earth resembles a great animal or rather inanimate vegetable, draws in aethereal breath for its daily refreshment and vital ferment and transpires again with gross exhalations. And, according to the conditions of all things living, ought to have its time of beginning, youth, old age, and perishing. (Ibid., f. 3.v)[2]

Perhaps as a result of stating this animistic conception of the earth so explicitly, Newton then abruptly dropped his discussion of the gravitational function of the ether and resumed a discussion of its vegetative function. Indeed, it is not clear in these sections of the manuscript that the two functions are really quite distinct in his mind. An extensive quotation from the "Vegetation" manuscript is in order here, to emphasize the partial fusion of the mechanical and the nonmechanical functions of the ether in Newton's thought during the transitional period of the early 1670s.

> [This is the subtle spirit which searches the most hidden recesses of all grosser matter, which enters their smallest pores and divides them more subtly than any other material power whatever. (Not after the way of common menstruums by rending them violently asunder, etc.) This is nature's universal agent, her secret fire, the only ferment and principle of all vegetation. The material soul of all matter, which, being constantly inspired from above, pervades and concretes with it into one form; and then, if incited by a gentle heat, actuates and enlivens it; but so tender and subtle is it withal as to vanish at the least excess and (having once begun to act) to cease acting forever and congeal in the matter at the defect of heat, unless it receive new life from a fresh ferment. And thus perhaps a great part if not all the moles of sensible matter is nothing but aether congealed and interwoven into textures whose life depends on that part of it which is in a middle state, not wholly distinct and loose from it like the aether in which it swims as in a fluid, nor wholly joined and compacted together with it under one

> form; in some degree condensed, united to it yet remaining of a much more rare, tender, and subtle disposition; and so this seems to be the principle of its acting, to resolve the body and be mutually condensed by it, and so mix under one form, being of one root, and grow together till the compositum attain the same state which the body had before solution. Hence 1. the earth needs a constant fresh supply of aether. 2. Bodies are subtiliated by solution]. (Ibid., ff. 3v–4r.)

"Moles" are ether condensed, yet the rarer, tenderer, and more subtle parts of the ether are those that contribute life. The rarer parts are thus more active than the general ether in which they swim "as in a fluid," yet they are "united" to it. Thus were the gravitational and vegetative functions conjoined so that the mechanical gravitational ether became the bearer of the life-giving spirit.

Newton, however, set brackets around that entire passage, an act that often may be taken to mean a reconsideration on his part of the bracketed material, perhaps with an eye to cancellation. Certainly, in this instance, he seems to have made a brief effort to differentiate again the mechanical (passive) and vegetative (active) ethereal functions, for immediately after the bracketed passage he began a new paragraph by saying: "Note that 'tis probable the aether is but a vehicle to some more active spirit." That statement seems to place the main body of the ether — presumably that part responsible for gravitation — firmly in the category of mechanical action again. Yet, if that was Newton's intention, he was not very succesful in his effort for, in the remainder of the paragraph, the vegetable spirit remains "entangled" in the ether, and generating and growing bodies imbibe ether as well as air and spirit, which seems again to link even the gravitational ether to the process of vegetation and to give it a more than mechanical function.

> Note that 'tis probable the aether is but a vehicle to some more active spirit. The bodies may be concreted of both together; they may imbibe aether as well as air in generation, and in that aether the spirit is entangled. This spirit is perhaps the body of light because both have a prodigious active principle, both are perpetual workers. (Ibid., f. 4r.)

By 1675 Newton's thoughts had matured to the extent that he was willing to launch them into the world in "An Hypothesis explaining the Properties of Light, discoursed of in my severall Papers." When one compares the "Hypothesis," which he sent to the Royal Society, with the private manuscript on "Vegetation," one sees that two notable shifts have taken place.

First, the gravitational principle has been further spiritualized. By 1675 it was *not* the relatively phlegmatic bulk of the ether that was responsible for gravitation, but a special, thin part of it, something especially subtle that was similar to the vital spirit in air. Though still material, Newton's ether had become extraordinarily complex and was filled with vapors, exhalations, and effluvia. After treating electrical matter as ethereal and suggesting condensation of ethereal electrical matter as the cause of what we call electrical attraction, Newton made a parallel suggestion for gravitational attraction:

> [S]o may the gravitating attraction of the Earth be caused by the continuall condensation of some other such like aethereall Spirit, not of the main body of flegmatic aether, but of something very thinly & subtily diffused through it, perhaps of an unctuous or Gummy, tenacious & Springy nature, and bearing much the same relation to aether, wch the vitall aereall Spirit requisite for the conservation of flame & vitall notions ... does to Air. (*Correspondence*, 1:365.)

So a state of corporeality even finer than that of the main body of the ether had definitely been assigned to the gravitational "spirit" by 1675. In addition, the operation of this "aethereall Spirit" by mechanical impact had been further eroded by a stronger emphasis on its unctuous, gummy, tenacious, and springy qualities. And, even more significantly, the gravitational function and the vegetative function of the ether had been even more closely fused than in the "Vegetation" manuscript, for the passage just quoted continued, without a break, with a discourse on what is transparently the vegetative function, even though the topic as Newton introduced it had been concerned with gravitation only.

> For if such an aethereall Spirit may be condensed in fermenting or burning bodies, {or otherwise inspissated in ye pores of ye earth to a tender matter wch may be as it were ye succus nutritious of ye earth or primary substance out of wch things generable grow} [or otherwise coagulated, in the pores of the earth and water, into some kind of humid active matter for the continuall uses of nature, adhereing to the sides of those pores after the manner that vapours condense on the sides of a Vessell subtily set]; the vast body of the Earth, wch may be every where to the very center in perpetuall working, may continually condense so much of this Spirit as to cause it from above to descend with great celerity for a supply. (Ibid., 1:365–66; curly and square brackets inserted by the editor.)

With the gravitational and vegetative functions thus joined, the gravitational ether fully shared in the activity of the vegetable spirit. The gravitational particles were not only thin and subtle; when condensed they became a "tender matter," the "succus nutritious," the "primary substance out of wch things generable grow," or a "humid active matter." By conflation with the active vegetable spirit, gravity had become an active principle.

Yet the general notion of the gravitational ether was still essentially the same as it had been for ten years, since the original entry in his student notebook, for Newton explained, again without a break, how that descent of the ether "with great celerity" contrived to bring bodies down with it.

> In which descent it may beare downe with it the bodyes it pervades with force proportionall to the superficies of all their parts it acts upon; nature makeing a circulation by the slow ascent of as much matter out of the bowells of the Earth in an aereall forme wch for a time constitutes the atmosphere, but being continually boyed up by the new Air, Exhalations, & Vapours riseing underneath, at length, (Some part of the vapours wch returne in rain excepted) vanishes againe into the ethereall Spaces, & there perhaps in time relents, & is attenuated into its first principle. (Ibid., 366.)

So the movement up of air and the movement down of ether continue in this still earthbound system. "For nature is a perpetual circulatory worker.... Some things to ascend & make the upper terrestrial juices, Rivers and the Atmosphere; & *by consequence* others to descend for a Requitall to the former" (ibid., my italics). In the "Hypothesis" of 1675 we find at first only a slightly refined restatement of the "crowding" of the ether by the air that we saw in the "Vegetation" manuscript. The motion is like that of the ether in its old mechanical days: the ether descends to the earth, circulates through the bowels of the earth, condenses, and reemerges as "grosser" forms of matter that then ascend in a great circulatory pattern. The larger particles of matter, constituted as they are only of condensed ether, are "attenuated" again in the ethereal region just beyond the atmosphere, then descend once again as ether. However, even though the movement is the same, the association of the gravitational function with the vegetative function has done its work. Gravity no longer arises from "the maine body of flegmatic aether" but from a portion of it that is thin, subtle, active, and spiritual, identical in fact with the vegetable spirit.

But with increased activity, gravity also suddenly moved toward universality, and that is the second important difference to be found in Newton's "Hypothesis." Appearing in the published paper almost as an afterthought, at the end of his description of the great circulation down to the earth, through it, and back up to the ethereal regions, the operations of the gravitation/vegetable "Spirit" were expanded to include the entire solar system.

> And as the Earth, so perhaps may the Sun imbibe this Spirit copiously to conserve his Shineing, & keep the Planets from recedeing further from him. And they that will, may also suppose, that this Spirit affords or carrys with it thither the solary fewell & materiall Principle of Light; And that the vast aethereall Spaces between us, & the stars are for a sufficient repository for this food of the Sunn & Planets. (Ibid.)[3]

The concept there enunciated bears little resemblance to the "real" Newton's "real" universal gravity, operating by mathematical law and applicable to all bodies in the universe. In that passage there has only been an equivocal extension of the animistic concept of the "Vegetation" manuscript to include the sun and the other planets. There the earth only, that "great animal or rather inanimate vegetable," drew in its "aethereal breath for its daily refreshment and vital ferment." By 1675, what had been food only for the earth had become food for the entire solar system, while the ethereal "repository" had been moved out of its place just beyond earth's atmosphere to the "vast aethereal Spaces" between the solar system and the fixed stars.

Newton's speculative ethereal system was enormously expanded in its scale of application when he thus extended it to the sun and other planets, but in no way did that expansion affect the system's mode of operation, for both gravitational and vegetative functions were still attributed to the "Spirit." The "Spirit" had the stated gravitational function of keeping the planets in their orbits: the sun imbibed the "Spirit" to "keep the Planets from recedeing further from him." And, since the gravitational function of the ethereal spirit had just been mentioned with respect to the earth in the "Hypothesis," by a reasonable analogy the "Spirit" would also bear bodies down to the surfaces of the other planets, though Newton made no explicit statement on that point in 1675. The vegetative function of the "Spirit," on the other hand, is readily apparent in other phrases: this spirit provides "food" and "fewell." It furthermore carries with it, or actually "affords," the "materiall Principle of Light," a sharp and unmistakable echo of Newton's identification of the vegetable spirit with "the body of light" in the "Vegetation" manuscript.

The entire conceptual structure of Newton's ethereal hypothesis of 1675 is so closely affiliated with that of the alchemical "Vegetation" manuscript that one would hesitate to suggest a relationship between it and his later concept of universal gravitation, had not Newton himself claimed in 1686 those very words quoted from the "Hypothesis" as evidence that he had the concept of universal gravity in 1675. The occasion for Newton's claim was Hooke's attempt to establish his own priority, and Newton in white anger looked through his manuscripts until he found the exact passage, then begged Halley to search for the Royal Society's copy (Newton to Halley, 20 June 1686, *Correspondence*, 2:435–41, esp. 439).

After quoting to Halley the section of the "Hypothesis" beginning "And so the Earth, so perhaps may the Sun,..." Newton said:

> In these & ye foregoing words [i.e., the discussion of his earthbound system] you have ye common cause of gravity towards ye earth Sun & all the Planets, & that by this cause ye Planets are kept in their Orbs about ye Sun. And this is all ye Philosophy Mr Hook pretends I had from his letters some years after, the duplicate proportion only excepted. The preceding words contein ye cause of ye phaenomena of gravity as we find it on ye surface of the earth without any regard to ye various distances from ye center: For at first I designed to write of nothing more. Afterwards, as my manuscript shews, I interlined ye words above cited relating to ye heavens, & in so short & transitory an interlined hint of things, the expression of ye proportion may well be excused. But if you consider ye nature of ye Hypothesis you'l find that gravity decreases upward & can be no other from ye superficies of ye Planet then reciprocally duplicate of ye distance from the center, but downwards that proportion does not hold. (Ibid., 2:439–40.)

The "duplicate proportion" was not an insignificant aspect of Newton's gravity a decade after 1675, when he wrote the *Principia*, but did he really have it in mind when he wrote the "Hypothesis"? Perhaps he did. The weight of the evidence seems to indicate that he had formulated the inverse square relationship with respect to objects gravitating toward the earth in the 1660s, as he later claimed to have done (Rosenfeld 1969; Aiton 1969; Whiteside 1964, 1970; Westfall 1983). But he can hardly have applied it universally even by 1675, for the papers show that he did not arrive at a concept of the mutual attraction between gravitating bodies until he was actually writing the *Principia* in the 1680s (Westfall 1980; Cohen 1980). What he did certainly have in 1675, or at least believed himself to have, was the *cause* of gravity. Later, he was to say that the cause of gravity was what he did not pretend to know, but in 1686, looking back to his ethereal hypothesis of 1675, he was eager to claim that knowledge: the words of the "Hypothesis" 'contein ye cause of ye phenomena of gravity as we find it on ye surface of the earth" and they also contain "ye common cause of gravity towards ye Earth Sun & all the Planets."

Conclusion

Newton continued to mention vortices quite casually in his correspondence of the early 1680s, as if some sort of ethereal mechanism were still an unquestioned assumption.[4] But about 1684 everything changed. The laws of planetary motion he had derived in *De motu* indicated little or no retardation of a planet's motion by the medium through which it moved. Newton had already been involved in the process of making the ethereal particles responsible for gravity finer and more subtle and active, as we have seen. In 1684 he saw that he would have to abandon that process and indeed abandon the corporeal ether altogether. In an incomplete treatise that he probably intended at one time to be a companion piece to *De motu*, he spoke explicitly of that new insight.

> For if the aether were a corporeal fluid entirely without vacuous pores, however subtle its parts are made by division, it would be as dense as any other fluid, and it would yield to the motion of bodies through it with no less sluggishness.... (*Unpublished Papers*, 147 [Latin, 113])[5]

And in the same paragraph, a little earlier:

> And so if there were any aerial or aethereal space of such a kind that it yielded without any resistance to the motion of comets or any other projectiles I should believe that it was utterly void. (Ibid., 146 [Latin, 112–13])

That there is such a space is what he had just demonstrated.

In an alchemical manuscript of the same period we find similar evidence of rethinking. Newton started to write something regarding the subtlety of the alchemical spirit but canceled that and wrote instead something about "the penetrating force of spirit and fixed force of body" (Keynes MS. 28, f. 6v). "The fixed force of body" is what he was currently attempting to define in revisions to *De motu* as the inertial powers of matter (Westfall 1980, 1983). "The penetrating force of spirit," on the other hand, was not a phrase he could have applied to any created spirit. All created spirits were to some degree corporeal and so had that element of corporeal "subtility" that he had marked out in this manuscript. "The penetrating force of spirit" was a phrase he could only have applied to the Deity himself — the one uncreated spirit. In this manuscript Newton seems to be talking only about the vegetative function of the spirit, but since the gravitational function had been assimilated to the vegetative during the 1670s, one must assume that the concept of the ethereal spirit in its entirety was involved in this final act of spiritualization.

Afterward, Newton sometimes still spoke of a "medium" or even an ether, but its resistance was null or very small and whether it was corporeal or incorporeal he would not or could not say. There is considerable evidence that he thought sometimes that gravity must be a direct operation of the will of God and that God's literal omnipresence constituted the incorporeal medium in which bodies moved. But there is also considerable evidence that he was uneasy with that solution and that he continued to search for a corporeal agent for both gravity and vegetation. The two functions of his ether had become inextricably joined in the 1670s and, though he never found a fully satisfactory explanation for either one, gravity and the vegetable spirit of alchemy both remained to the end "active principles" in his mind.

Notes

1. Henry More to Descartes, 23 July 1649, in Descartes, *Correspondance*, 170–81, esp. 176–79; also in Descartes, *Oeuvres*, 5:376–83, esp. 381: "Si naturaliter igitur moueretер materia, nec sol, nec coelum, nec terra esset, nec vortices ulli, nec heterogenum quicquam, siue imaginable, in rerum natura."
2. Newton probably derived this concept from the Stoics. Cf. Seneca, *Quaestiones naturales* 6:16.1: "Non esse terram sine spiritu palam est, non tantum illo dico quo se tenet ac partes sui iungit, qui inest etiam saxis mortuisque corporibus, sed illo dico vitali et vegeto et alente omnia." I am indebted to David E. Hahm for this reference to Seneca.
3. Newton later told Halley that he had "interlined" this material in his own draft of the letter sent to the Royal Society, confirming the impression that his expansion of the operations of the "Spirit" to the solar system was something of an afterthought. Cf. *Correspondance*, 2:435–41 (Newton to Halley, 29 June 1686), esp. 439–40 (and cf. also below).
4. Cohen 1980, 114–20, provides a convenient recent summary of Newton's changing ideas on the ether, subtle media, and vortices.
5. The quotation is from Portsmouth Collection, MS. Add. 4003, "De gravitatione et aequipondio fluidorum," published and translated in *Unpublished Papers*, 90–156. The handwriting of this manuscript is indistinguishable from that of the *De motu* MS., dated by Whiteside at 1684/85, and from that of Newton's alchemical "Commentarium" in Keynes Collection, MS. 28 (quoted below), independently dated at 1864 by the present author.

References

Printed Works of Newton

The Correspondence of Isaac Newton, 1959–1978. Ed. H. W. Turnbull et al., 7 vols. Cambridge: Cambridge University Press.

Opticks, 1952. Based on 4th ed. New York: Dover.

Unpublished Scientific Papers of Isaac Newton, 1962. Selected, edited, and translated by A. R. Hall and M. Boas Hall. Cambridge: Cambridge University Press.

Manuscript Works of Newton

Burndy Collection, Smithsonian Institution, Washington, D.C.
Keynes Collection, Kings College, Cambridge.
Yahuda Papers, Jewish National and University Library, Jerusalem.

Other References

Aiton, E., 1969. "Newton's Aether-Stream Hypothesis and the Inverse Square Law of Gravitation," *Annals of Science* 25:255–60.

Burtt, E. A., 1954. *The Metaphysical Foundations of Modern Science*, repr. of 2nd rev. ed. Garden City, N.Y.: Doubleday.

Cassirer, E., 1953. *The Platonic Renaissance in England*, trans. J. P. Pettegrove. Austin, Texas: University of Texas Press.

Cohen, I. B., 1980. *The Newtonian Revolution*. Cambridge: Cambridge University Press.

Descartes, René, 1953. *Correspondance avec Arnauld et Morus*, with introduction and notes by Geneviève Lewis. Paris: Librairie Philosophique J. Vrin.

—, 1964–1974. *Oeuvres*, published by C. Adam and P. Tannery, 11 vols. Paris: Librairie Philosophique J. Vrin.

Dobbs, B. J. T., 1975. *The Foundations of Newton's Alchemy, or "The Hunting of the Greene Lyon."* Cambridge: Cambridge University Press.

—, 1979. "Newton's Copy of *Secrets Reveal'd* and the Regimens of the Work," *Ambix* 26:145–69.

—, 1982. "Newton's Alchemy and his Theory of Matter," *Isis* 73:511–28.

—, 1982. "Newton's 'Clavis': New Evidence on its Dating and Significance," *Ambix* 29:190–202.

—, 1984. "Alchemische Kosmogonie und arianische Theologie bei Isaac Newton," trans. Ch. Meinel, in "16. Wolfenbutteler Symposium: Die Alchemie in der Europäischen Kultur- und Wissenschaftsgeschichte der frühen Neuzeit... vom 2. bis 5. April 1984." *Wolfenbutteler Forschungen*.

—, 1985. "Newton and Stoicism," in *Southern Journal of Philosophy*.

—, 1988. "Newton's *Commentary* on *The Emerald Tablet* of Hermes Trismegistus: Its Scientific and Theological Significance," in *Hermeticism and the Renaissance. Intellectual History and the Occult in Early*

Europe, edited by Ingrid Merkel and Allen G. Debus. Washington, D.C.: The Folger Shakespeare Library; London/Toronto: Associated University Presses, 182–91.

Emerton, N. E., 1984. *The Scientific Reinterpretation of Form*. Ithaca, N.Y.: Cornell University Press.

Grant, E., 1981. *Much Ado about Nothing. Theories of Space and Vacuum from the Middle Ages to the Scientific Revolution*. Cambridge: Cambridge University Press.

Hager, P., 1982. "Chrysippus' Theory of the Pneuma," *Prudentia* 14:97–108.

Hahm, D. E., 1977. *The Origins of Stoic Cosmology*. Columbus, Ohio: Ohio State University Press.

Hutin, S., 1966. *Henry More. Essai sur les doctrines theosophiques chez les Platoniciens de Cambridge*. Hildesheim: George Olms.

Lapidge, M., 1973. "Archai and Stoicheia: A Problem in Stoic Cosmology," *Phronesis* 18:240–78.

McGuire, J. E., 1968. "Force, Active Principles, and Newton's Invisible Realm," *Ambix* 15:154–208.

—, 1977. "Neoplatonism and Active Principles: Newton and the *Corpus Hermeticum*," in *Hermeticism and the Scientific Revolution*, 93–142, Papers read at a Clark Library Seminar, 9 March, 1974, by Robert S. Westman and J. E. McGuire. Los Angeles: University of California, Los Angeles.

McGuire, J. E., and M. Tamny, 1983. *Certain Philosophical Questions: Newton's Trinity Notebook*. Cambridge: Cambridge University Press.

McMullin, E., 1978. *Newton on Matter and Activity*. Notre Dame, Ind.: Notre Dame University Press.

More, Henry, 1659. *The Immortality of the Soul* ... , London: printed by J. Flesher for William Morden, Bookseller in Cambridge.

Pagel, W., 1958. *Paracelsus: An Introduction to Philosophical Medicine in the Era of the Renaissance*. Basel: Karger.

Rosenfeld, L., 1969. "Newton's Views on Aether and Gravitation," *Archive for History of Exact Sciences* 6:29–37.

Taylor, F. Sherwood, 1953. "The Idea of the Quintessence," in: *Science, Medicine, and History. Essays on the Evolution of Scientific Thought and Medical Practice Written in Honour of Charles Singer* 1:247–65. Collected and edited by E. Ashworth Underwood, 2 vols. London: Oxford University Press.

Westfall, R. S., 1971. *Force in Newton's Physics. The Science of Dynamics in the Seventeenth Century*. London: Macdonald; New York: American Elsevier.

—, 1980. *Never at Rest. A Biography of Isaac Newton.* Cámbridge: Cambridge University Press.

—, 1983. "Newton's Development of the *Principia*," in *Springs of Scientific Creativity. Essays on Founders of Modern Science*, ed. R. Aris *et al.*, Minneapolis, Minn.: University of Minnesota Press.

Whiteside, D. T., 1964. "Newton's Early Thoughts on Planetary Motion: A Fresh Look," *British Journal for the History of Science* 2:117–37.

—, 1970. "Before the *Principia*: The Maturing of Newton's Thoughts on Dynamical Astronomy, 1164–1684," *Journal for the History of Astronomy* 1:5–19.

Isaac Newton: Theologian

RICHARD S. WESTFALL

I run some danger, I am aware, in offering a paper on Isaac Newton theologian, as part of a colloquium that bears the title *The Other Newton*. Can anyone possibly entertain the notion that an exposition of Newton's interest in theology can pass as a new revelation, in the way that Dobbs' work on his endeavor in alchemy has? Obviously not. And if only to reassure the reader, let me begin with a reminder of familiar themes that we all know were prominent in Newton's thought.

"When I wrote my treatise about our Systeme," he began a letter written on 10 December 1692 to the theologian Richard Bentley — surely one of the best-known sentences that Newton ever composed — "I had an eye upon such Principles as might work wth considering men for the beleife of a Deity..." (*Correspondence*, 3:233). From this point of view, science appeared as the handmaiden of religion; its purpose was not to manipulate nature for the material benefit of mankind, but to demonstrate the existence of the Creator. The "main Business of natural Philosophy," Newton asserted elsewhere (and let me insist on the significance of the adjective "main"), "is to argue from Phaenomena without feigning Hypotheses, and to deduce Causes from Effects, till we come to the very first Cause, which certainly is not mechanical..." Every step along the way brings us closer to the goal "and on that account is to be highly valued" (*Opticks*, 369–70). Again I want to insist on the significance of the phrase, "on that account."

Newton was always the physicist and, not surprisingly, it was the structure of the cosmos that demonstrated the existence of God to him. Let us assume, he argued in the letter to Bentley quoted above, that in the beginning, matter

E. Ullmann-Margalit (ed.) The Scientific Enterprise, 223–239.

was evenly diffused through the universe. If the universe were finite, mutual gravitation would have caused all matter to congregate together in a single body. If, on the other hand, the universe were infinite, matter would have congregated in a number of bodies held in equilibrium by their mutual attractions, and if all matter were luminous, we could explain the system of the sun and stars by natural causes alone. But all matter is not luminous, and by the operation of natural causes alone we cannot explain the separation of luminous from nonluminous matter. For this we have to call upon an intelligent Creator. Moreover, the sun is at the center of our system. If chance, that is, the blind operation of natural causes, had formed our system, the central body could just as well have been without light and heat; it required a God to understand the importance of making the central body a sun. Why is there only one sun in our system? God saw that one was sufficient. Moreover, all of the planets move in the same direction in the same plane, an ordered whole, as we plainly see when we compare the planetary system with the disorder of comets.

> To make this systeme therefore wth all its motions, required a Cause wch understood & compared together the quantities of matter in ye several bodies of ye Sun & Planets & ye gravitating powers resulting from thence, the several distances of the primary Planets from ye Sun & secondary ones from Saturn Jupiter & ye earth, & ye velocities wth wch these Planets could revolve at those distances about those quantities of matter in ye central bodies. And to compare & adjust all these things together in so great a variety of bodies argues that cause to be not blind & fortuitous, but very well skilled in Mechanicks & Geometry (*Correspondence*, 3:235).

In the *Principia*, Newton implicitly presented a similar argument. The first book of the *Principia* consistently examined the operation of two basic force laws, forces that vary inversely as the square of the distance, and forces that vary directly as the distance. Under these two laws, and under these two laws alone, bodies will revolve in conical orbits around central attracting bodies (located at the focus in the case of the inverse square attraction and at the center in the case of the direct distance one). Under these two laws, and, among plausible ones, under these two alone, orbits will remain stable in space with the line of apsides pointing always in the same direction relative to fixed stars. Under these two laws, and under these two laws alone, composite bodies made up of attracting particles will attract according to the same law as their components. The law of variation in proportion to distance is not suitable to an ordered universe, however. In an infinite universe, as Newton believed ours to be, such a law would entail infinite forces, infinite accelerations, and infinite velocities, which are physically absurd. The ultimate demonstration of the wisdom of God lies in the fact that he did in fact employ in the creation the only law suitable for the rationally ordered universe. God, we might say, had shown himself to be the equal of Newton in his understanding of the laws of mechanics. Or rather, let us say that only a scientist like Newton could fully appreciate the wisdom of God.

Not only are these arguments of Newton well known, but we know as well how thoroughly typical of his age they were. In the twentieth century, we tend to accept the notion of fundamental and unavoidable conflict between science and religion. Such was by no means the case in the seventeenth century. Though a small number of men, who appear to me to have stood on the outer fringes of thought, assumed that science and religion were in conflict, the majority, and especially the majority of scientists, and more especially still the majority of scientists in Britain, thought in terms of harmony rather than conflict between the two realms of knowledge and activity. The works of God magnify the glory of God. The works they had studied. Glory they had found, and from one end of the century to the other they raised their hymn of praise to its author. Robert Boyle, a most prolific author, sounded the theme in virtually every one of his many books and, toward the end of his life, in case the message had failed to be heard, he formulated his definitive statement of it in *The Christian Virtuoso*, a title that we can translate without serious inaccuracy into more familiar language as "The Christian Scientist". John Ray, the leading naturalist of the age, stated the same idea in his book *The Wisdom of God Manifested in the Works of the Creation.* There were so many other versions of a similar outlook that it would be difficult to list all of them. When Newton found demonstrative evidence for the existence of God in the structure of the cosmos, his argument may have been unique in the type of evidence it cited, but in every other respect it repeated attitudes that had been expressed dozens, perhaps hundreds of times, in his age alone, before he put them on paper.

Thus it will be part of my argument that this theme represents the traditional element in Newton's religion. It is the most visible part of his religion, and until recently it was virtually the only aspect of it that was known. If we concentrate on these arguments alone, they are apt to mislead us, for two different reasons. In the first place, we may make the mistake of taking the arguments at face value. But Newton and the other scientists did not find God in nature. Quite the contrary, they imposed God upon nature. That is, the arguments did not so much derive from the study of nature as descend from the long tradition of Christianity in Western Europe. Consider Newton's letter to Bentley cited above. If God created an ordered universe, whence arose the disorder of the comets, to which Newton himself referred? By what criterion did Newton determine that the order of the planetary system was more typical of the whole than the disorder of the comets? If it was the purpose of God to reveal his wisdom by making the planets move in the same plane, we need to note that he bungled egregiously. The planetary planes are inclined to each other by as much as seven degrees. What eighteenth-century artisan, building an orrery, would have tolerated an error that large? That is, Newton's arguments reveal above all a determination to find God in nature. They were the deposit of centuries of Christianity in the West, an inherited piety, that part of Newton's religion not yet disturbed by the rise of modern science. "I don't know what I may seem to the world," he is reported to have said near the end of his life, "but, as to myself, I seem to have been only like a boy playing on the sea shore, and diverting myself in now and then finding a smoother pebble or a prettier shell than ordinary, whilst the great ocean of truth lay all undiscovered before me."[1] A similar figure appeared among the papers of Robert Boyle, who compared the student of nature to a man traveling down a great river. The farther he goes, the more the shores recede on either side until, when he reaches the river's mouth, he finds himself before the boundless sea.[2] The Christian Virtuoso of the seventeenth century was like the boy or man on the shore confronting the ocean of truth. His attitude resembled the conviction of medieval philosophers, an earlier expression of the tradition in which Newton stood, that there are truths that are, by their nature, beyond the capacity of the human mind.

The second reason why Newton's arguments for the existence of God are apt to mislead us is the possibility of mistaking his manifest piety for the whole of his religion. In fact the story was far more complicated. It was more complicated with all of the English scientists. When we read only one or two of their refutations of atheism, we may find them impressive testimony, but by the time we read the tenth repetition of the same argument, we begin to sense some uneasiness behind it. Boyle offers a prime example. After a lifetime devoted to the refutation of atheism, he left provision in his will to endow a series of public lectures. What were the lectures supposed to do? Refute atheism some more. When during the previous fifteen hundred years had that appeared necessary? Boyle was aware that the ground was shifting under the traditional foundations of Christianity.

Newton was also aware of that fact, but his reaction differed from Boyle's. Instead of trying to shore up the established foundations, Newton attempted to make the central structure secure by abandoning its faulty members. Lest I be misunderstood, let me dispense with the figure of speech and state my proposition in more direct terms. I mean to say that Newton questioned orthodox theology and rejected those of its teachings that he found contrary to reason. In order to discuss this issue satisfactorily, we must make a distinction between religion and theology. Nothing that I may say is meant to question Newton's sincerity in the passages I quoted above. There can be no doubt that Newton believed in the existence of God. Merely to state his belief in these terms is grossly to undervalue it. His belief in God was a living faith that suffused his entire life and gave it meaning. Equally he was convinced that science is in harmony with religion. The God demonstrated to exist from his work in the creation is not, however, necessarily identical to the God of received Christianity. In the case of Newton, he certainly was not. At this point we begin to confront the "other" Newton, for the religious beliefs that he held differed radically from the conventional piety apparently presented in the published works, the piety we have taken until now as a typical manifestation of his religion.

We cannot effectively explore this aspect of Newton's religion through his published works. Newton was a man who feared controversy. In the matter of religion, he had good cause to fear controversy, for he had much to lose. While he was at Cambridge, the views that he came to hold during his stay there would have been grounds for his immediate dismissal, and after his move to the Mint in London, those same views made him ineligible, according to the law of the land, to hold a position in the government. Knowing this full well, he took care that his religious beliefs should not become matters of public knowledge. Newton was a compulsive writer, however, and he left behind a huge collection of private papers. I can speak properly of the "other" Newton, because the great bulk of his theological manuscripts became available for study only very recently, and they have a story to tell that differs from the one in his published works with which we are familiar.[3] I will not have to tell you that the body of manuscripts to which I refer, part of the Yahuda Papers, have come to rest in the Jewish National and University Library in Jerusalem and are the specific occasion of my participation in this colloquium. Although, as I understand, an increasing number of scholars are beginning to study them, there is, as yet, very little about them in print, and they remain largely unknown. Since this is the case, let me cast my account in the form of a description of their content, with an implicit interpretation embedded in the description.

The papers reveal that Newton began the serious study of theology in about 1670, when he was approaching the age of thirty.[4] Note that Newton began intensive theological study when he was a young man, at the very height of his powers. Contrary to what has sometimes been asserted, such study was not solely the occupation of his old age, although he did devote a great deal of time to it then as well. The papers do not offer any explanation of why he took up theology, but I am willing to hazard the suggestion that the statutes of Trinity College had something to do with it. The statutes ordained that the holders of fifty-eight of the college's sixty fellowships had to be ordained to the clergy of the Anglican church within seven years of incepting M.A., or face ejection. Newton was never one who took an obligation lightly. His seven years would have expired in 1675, and the approaching deadline would have been incentive enough. Whatever his motives may have been, suffice it to say that the papers establish beyond any doubt the fact that he did undertake vigorous theological studies in about 1670.

We have every reason to think that he was at that time entirely orthodox. In the recent past he had taken solemn oaths to that effect on three separate occasions. With each of his two degrees, in 1665 and 1668, he had sworn to three articles, one of which affirmed the faith of the Anglican church, and when he had accepted a fellowship in the College of the Holy and Undivided Trinity, he had sworn that he would uphold the one true religion, in a context that equated the one true religion with the doctrine of the Anglican church. Although he had not taken an oath with the Lucasian chair, it did embody a similar requirement; its statutes made heretical belief grounds for dismissal. Everything we know about Newton indicates that he would not have sworn falsely. He must have been orthodox at the time.

In accordance with his universal practice when he turned to a new topic of study, Newton began a systematic compilation of his reading notes, which would organize his new knowledge in an effective manner. He purchased a notebook in which he entered a series of headings under which he expected to gather the fruits of his study (Keynes MS. 2). The pages under some of the headings — "The Miracles of Christ," for example — remained blank. Such was not the case for "God the Son" and "Concerning the Trinity," which immediately became the foci of his interest. Initially, the Bible furnished his reading, and his intensive study of the Scriptures at this time furnished the foundation of the detailed knowledge of them that he carried through his life. From the Bible he advanced to the early Fathers of the Church. It is necessary to plunge into the manuscripts themselves in order fully to appreciate them. They reveal a vast program of reading, which led Newton through all of the important Fathers of Christianity.

They reveal as well a driving passion, which animated the study. What we meet in the manuscripts is not conventional piety assembling arguments designed to defend the tradition. Quite the contrary; what we meet is the passion of a rebel who had convinced himself that the received tradition was mistaken. "Mistaken" is far too mild a word. Newton convinced himself that the received tradition was a fraud perpetrated by evil men in the fourth century who, for their own selfish purposes, had willfully corrupted the entire heritage. Newton's determination to unmask this ancient crime, together with his study of alchemy, absorbed virtually all of his time for the following fifteen years before a visit from Edmund Halley started the investigation that resulted in the *Principia* and altered the tenor of his existence.

To be specific, Newton embraced Arianism, a fourth-century heresy which had struggled with Trinitarianism (as expounded especially by Athanasius) for the soul of Christianity. Arianism is similar to modern Unitarianism, insofar as it rejects the full divinity of Christ. It differs from Unitarianism in its belief that Christ was also not wholly human. The Arian Christ was a created intermediary between God and man. Newton hinted broadly at an Arian Christology in his theological notebook. In later manuscripts he did more than hint. Among them there is what appears to me his most explicit statement, a sheet from the period 1672–1675 with twelve propositions on the nature of Christ, which are wholly Arian in tone (Yahuda MS. 14, f. 25). From the position taken in the early 1670s Newton never retreated, although in one manuscript treatise he does appear to have gone further yet. In his old age, he was still composing Arian statements on the nature of Christ.

The position taken in the early 1670s nearly blighted Newton's career. He still faced the necessity of ordination. In an institution which had learned to pay scant attention to its governing statutes, requirements of ordination were one rule that was never broken. One can understand why. The colleges were filled with aspiring young men, and some no longer so young, patiently working — or better, waiting — their way up the ladder of seniority. They had nothing to gain and much to lose by tolerating the presence of anyone higher in the order of seniority who could be forced out for a failing they had no incentive to imitate themselves. By 1675, Newton had to accept ordination or resign his fellowship. As a result of his Arianism, he could no longer accept ordination.

There was a possible avenue of escape from his dilemma. Two of the sixty fellowships in Trinity were exempted from the requirement. In 1673, one of the two fell vacant, and Newton forthwith sought appointment to it. Robert Uvedale, a fellow one year his senior, also sought it. The rule of seniority was adamant, and the avenue of escape appeared to close. In a letter to Henry Oldenburg, the secretary of the Royal Society, written early in 1675, Newton mentioned that he would soon have to lay down his fellowship (*Correspondence*, 7:387). Perhaps Newton could have stayed on in Cambridge as the Lucasian Professor, although I do not know of another professor who was not associated with a college. The problem was not legality; the chair did not require a fellowship. The problem was the questions that would be asked. Why would anyone in his right mind, who intended to stay on at the university, give up a fellowship worth some sixty-five pounds per annum? Since ordination would obviously have been the sticking point, what would Newton's motives have to have been? Questions would have been asked. It would have been difficult indeed to keep his opinions secret, and once they were out, the professorship — and indeed his entire standing in acceptable society — would have been forfeited.

At the last moment, the cloud lifted. A royal dispensation exempted, not Isaac Newton, but the holder of the Lucasian Chair in perpetuity from any college requirement of ordination. The dispensation raises a number of questions which cannot be answered with assurance. Who arranged it? The most likely agent was Isaac Barrow, the Master of Trinity College, a man known to have had the king's ear. Why did he arrange it? Barrow was a man of unimpeachable orthodoxy, the author of a published defense of the doctrine of Trinity. I find it impossible to imagine him listening with sympathy to a plea of Arianism. I do not find it impossible that he could have accepted a plea of no vocation for the clergy. Barrow valued learning, and he understood the full dimensions of Newton's quality. He would have been anxious to have him continue in the college and the university. Whoever the agent and whatever his reasons, it is recorded fact that a dispensation granted in the spring of 1675 freed Newton from the requirement of ordination and allowed him to stay on in his academic sanctuary. And, in continuing silence, he continued to write theology.

Along with theology he early developed an interest in the prophecies. Newton composed his first interpretation of the Book of Revelation in the early 1670s, and he worked assiduously at expanding and revising it through the rest of the decade and into the early 1680s. His interest in the prophecies is well known. After his death, his heirs published his *Observations upon the Prophecies*. A work of surpassing tedium, which all but the tiniest handful have been spared the necessity to read, the *Observations* defy the reader to find a point in their meandering discussion. The published manuscript was a product of Newton's old age, deliberately made obscure in order to conceal his point. The reader of his papers has no problem in finding a point in his early interpretation. It embodied an adaptation of the standard Puritan interpretation, which hinged on the concept of the Great Apostasy. To Puritan exegetes, the Great Apostasy was Roman Catholicism. To Newton, the Great Apostasy was Trinitarianism. That is, his interpretation of the prophecies offered an alternative statement of his theological position. The plagues and the vials of wrath in the Book of Revelation were prophetic forecasts of the barbarian invasions of Europe, the Almighty's punishments on a stiff-necked people who had gone whoring after false gods.

The work contained an inner tension. On the one hand, it implied a chronology. In Newton's interpretation, Revelation contains repeated references under different figures to a period of 1,260 years immediately preceding the sounding of the seventh trumpet, which will announce the Second Coming of Christ and the Final Judgment. Newton was quite explicit in detailing the beginning of that period in 607, from which it follows that he placed the Final Judgment nearly two hundred years in the future. Not by any standard would one call that imminent. On the other hand, there is a sense of expectation in the work that is impossible to miss. The meaning of the prophecies is finally being revealed, and the Great Day must correspondingly be at hand. Intense passion accompanied the inner tension. "Idolators, Blasphemers & spiritual fornicators," Newton thundered at the Trinitarians in the silence of his study in Trinity. And, because only Scripture could adequately convey his fury, "Seducers waxing worse & worse, deceiving & being deceived — such as will not endure sound doctrine but after their own lusts heap to themselves teachers, having itching ears and turning away their ears from the truth unto fables" (Yahuda MS. 14, ff. 67–68; 1.2, f. 26). There is no doubt in my mind that Newton considered himself as one of the remnant persecuted by the dragon, one of the saints on whom the Beast made war, and he appears to have been expecting early release. Once again, the Great Day must be at hand. Against whom was the passion directed? There is also no doubt in my mind that it was directed at flaccid orthodox Cambridge all about him, the Cambridge from which Newton had largely withdrawn as he isolated himself within the fastness of his own study.

The interpretation of the prophecies implied another program of reading, indeed two more. In order to grasp their message, Newton needed to assure himself that he had their correct text. To this purpose he collated some twenty-five different Greek versions of the Book of Revelation to establish the true text, and he combed the Bible, of which he considered the prophecies to be the central books, to find confirming passages. The interpretation also implied an exact correlation between the prophetic text and the events of history as they later transpired. Although he read and used the works of his contemporary authorities, he had no intention of resting satisfied with anything less than original sources. For example, Revelation speaks of a silence of half an hour before the sounding of the first trumpet. In Newton's scheme, half an hour corresponded to seven and a half years, which ended in this case with the war between Theodosius and Maximus. In order to track the movements of Maximus in the early 380s, Newton called upon the testimony of Zosimus, Pacatus, Sulpitius Alexander (an eyewitness quoted by Gregory of Tours), the letters of Jerome and Ambrose, the *Annalium Boiorum*, and Gothofredus' commentary on the Theodosian Code. One brief passage on the period from Constantine through 380 A.D. cited Theodoret, Sozomen, Socrates (the historian), Ammianus, Claudian, Zosimus, Eusebius, Sigonius, Jerome, Eunapius, Libanus, Marcellinus, Victor, Gregory the Presbyter, Pacatus, Symmachus, Idatius, and Cassiodorus, plus the Theodosian Code, the Alexandrian Chronicle, and the letters of Ambrose (Yahuda MS. 1.4, ff. 33–42, 55–56). Do not mistake my meaning. No one, I believe, would confuse Newton's interpretation of the prophecies with great history. The point I wish to assert is that the interpretation did not rest on superficial research.

Repeatedly, the interpretation revealed itself as the work of the man who had recently composed *De methodis*, the definitive exposition of his fluxional calculus, and would soon write the *Principia*. His intention, he insisted, was to methodize the study of the prophecies. He set forth fifteen rules of interpretation — consistently rationalizing rules, such as always to assign only one meaning to one figure and always to prefer the simplest and most literal meaning — followed by a catalog of seventy figures and a section called the *Proof*, which cited the evidence that supported the meanings he had assigned to the seventy figures. Newton was convinced that there had been an accepted language of prophecy in the ancient world, which he could decipher. For the *Proof*, he combed the entire Bible plus the writings of Achmet, an Arabian authority on prophecies, and Artemidorus, the Hellenistic author on the interpretation of dreams, for corroborating instances. His stated goal was to free the interpretation of the prophecies from individual fancy and to reduce it to demonstration. Passages in his early treatise sound remarkably like passages on method in his letters to the Royal Society on optics, which date from the same period, and like the fourth Rule of Reasoning in Philosophy, which he inserted many years later at the head of Book III of the *Principia*. In its original version, he cast the interpretation in Propositions, though he later altered the word to Positions.

In the late 1670s, Newton embodied his interpretation of the prophecies in a history of the early church, a history which, inevitably, cast Athanasius in the role of villain. The history of the church, in turn, led into the well-known essay, "Paradoxical Questions concerning the morals & actions of Athanasius and his followers," one of the small number of Newton's theological manuscripts that have been available for a long time. In this period he also became interested in the rise of monasticism, which he associated with

Athanasius and Trinitarianism, and the frequent stories of early monks about their temptations and unclean thoughts, which he collected at some length, clearly fascinated him. Solemnly, he lectured the monks on how to cope with unchaste thoughts.

> For lust by being forcibly restrained & by struggling wth it is always inflamed. The way to be chast is not to contend & struggle with unchast thoughts but to decline them keep the mind imployed about other things: for he that's always thinking of chastity will be always thinking of weomen & every contest wth unchast thoughts will leave such impressions upon the mind as shall make those thoughts apt to return more frequently. (*Paradoxical Questions*)

Whatever the value of the advice, it came some 1,300 years too late to help the libidinous monks in the Egyptian desert.

The interpretation of the prophecies acquired a new dimension in the early 1680s, when Newton began to study Judaism. Again, we find a typically Newtonian exercise, which saw him plunge into the talmudic scholars, Josephus, Philo, and Maimonides. He concluded that Jewish rites and the details of Jewish worship had furnished prophetic figures, a theme that survived prominently in the published *Observations*. The study of Judaism led him to investigate the plan of the original Temple, a question that took on a life of its own as questions frequently did with Newton. He concluded that Chapters 40 through 43 of Ezekiel contained the best description of the Temple, and he reconstructed the text of those chapters, using the text to draw a plan and using the inherent necessities of the plan to amend the text. The Temple, in turn, raised the further question of the exact length of the sacred cubit of the Jews, which Ezekiel had used to measure it, and after extensive investigation Newton concluded that it had been more than 25.2 and less that 26.25 Roman inches (Babson MS. 434; Yahuda MS. 2.4, f. 40).

About this time, that is, in the early 1680s, Newton's interpretation began to suppress the passion that marked the early version and to shift toward the colorless chronology of the early church that characterized the version finally published after his death. The Great Apostasy, the concept that offered the key to the whole, became indistinguishable in his exposition from Roman Catholicism. No single word that I have found explains the change. I will offer the speculation that the rising specter of James Stuart and the threat of a Catholic succession led a man who hated Catholicism passionately to seek common ground with English Protestantism and to suppress those features of a work, albeit unpublished and unknown, which would have alienated him from other non-Catholics.

In the early 1680s Newton also began to compose a new theological work to which, in the least chaotic of its manuscript remains, he gave the name *Theologiae gentilis origines philosophicae* (Yahuda MS. 16.2). It was easily the most important theological treatise he ever wrote or, better, worked on, and traces of its continuing presence in his thought showed up through the rest of his life.

The central concept of the *Origines* asserted that all the ancient peoples had worshiped twelve gods, the same twelve gods which were associated with the seven planets, the four elements, and the quintessence. A second concept, not fully integrated with the first, claimed that the twelve gods were their deified ancestors, that is, Noah, his children, and his grandchildren, from whom all mankind had descended. The religion of twelve gods was not the earliest religion; Noah and his family themselves had worshiped the one true God, the Creator of the universe. However, there is an innate tendency toward superstitution and idolatry in man which led him to corrupt the true religion. The Egyptians had led the way in this; the other ancient peoples had learned their idolatry from Egypt. Each people had appropriated the religion to their own history, giving to the twelve gods the names their own records assigned to their ancestors, but the perceptive student of their chronicles could discern the common originals. God had made repeated efforts to lead mankind back to the true religion. Despairing of humanity as a whole, he had chosen the Jewish people and had sent Moses and the prophets to reclaim them. Despairing of the Jews, he had sent Jesus Christ to the Gentiles. Jesus had come to recall men to the true religion, which rested on two commandments, to love God and to love one's neighbor. He had added nothing to that religion, Newton insisted; he had come only to call mankind back to its original worship.

We must be careful not to misinterpret the *Origines*. We see references to Noah and the biblical account, quaint themes of seventeenth-century scholarship. We smile indulgently when we realize that even the great Newton accepted such childish notions, and we run the risk of missing the radical thrust that the quaint themes concealed. To Samuel Bochart, the Huguenot scholar on whom Newton drew extensively, Noah was literally Noah. That is, Noah, in the Hebrew original, of course, was the true name of the common forebear of mankind. To Newton, in contrast, Noah was merely the name most familiar to readers in Christian Europe. To Newton, Egypt rather than Israel was basic. Noah's son was more truly Ammon than Ham. Similarly, Newton treated the historical books of the Old Testament as nothing more than the historical records of the Jewish people, records no more authoritative than those of other peoples to which he compared them.

Above all, the *Origines* deflated the role of Christ in human history. Christ came to call mankind back to the one true religion, and to that religion he added nothing. Its two basic commandments, Newton insisted, "always have & always will be the duty of all nations & The coming of Jesus Christ has made no alteration in them" (Keynes MS. 3, p. 35). That is, Christ did not signal a new dispensation and a new religion; he merely recalled mankind to an old one. Trinitarianism, when in its turn it corrupted the restored worship once more, repeated the idolatry of earlier ages by worshiping a man as god. Athanasius, the principal author of Trinitarianism, even repeated the original nefarious role of Egypt in the formation of superstition. Along with its deflation of Christ, the *Origines* likewise suggested that the one true religion is known to mankind from the study of nature. Newton found evidence that everywhere the original inhabitants of the earth had worshipped in similar temples, *prytanea* as he called them, similar in plan to the Jewish Temple. The Roman temples of Vesta offered perhaps the most familiar version of such temples, which embodied a representation of the universe.

> The whole heavens they recconed to be ye true & real temple of God & therefore that a Prytanaeum might deserve y[e] name of his Temple they framed it so as in the fittest manner to represent the whole systeme of the heavens. A point of religions then w[ch] nothing can be more rational. (Yehuda MS. 41, ff. 6–7)[5]

We might note that the temples, with a fire in the center and representations of the planets around it, offered a heliocentric picture of the universe. The decay of true philosophy had accompanied the decay of true religion. Meanwhile, the *Origines* asserted that by the proper study of nature men could recognize their Creator and their duties toward him. Christ had affirmed no more.

Newton believed in the concept of divine revelation; as with so much, however, the concept carried a new meaning for him. To him the Bible was not the revelation of mysteries beyond reason unto life eternal. As an Arian he did not accept this position. I have indicated that he did not consider the historical books of the Old Testament to have been divinely revealed. Prophecy was the central element of revelation — prophecy, whereby God, through its fulfillment, demonstrated his dominion over history.

The *Origines* meanwhile appears to me to have gone beyond Arianism. Although it differed in its attitude from the deists' open hostility toward Christianity, its content was remarkably similar to their works, and I am prepared to argue that it must be seen as the first of the deist tracts, even if it remained unknown to any significant number and probably to everyone. As I have indicated, the *Origines* continued to appear important to Newton. Shortly after he composed the original manuscripts in which it survives, when he turned to the *Principia*, he chose to begin the first version of its final book with a passage from the *Origines*, though it did not survive in the published work.[6] Again he drew upon it for revisions of Book III in the 1690s, revisions which he also suppressed. The strange passage about Noah and his children that concludes the Thirty-first Query to the *Opticks* derived from it, and the "General Scholium" which he appended to the second edition of the *Principia* contained two footnotes based on the *Origines*. In his old age, he sanitized it thoroughly, and that manuscript was published after his death as the *Chronology of Ancient Kingdoms Amended*, a work that rivals the similarly sanitized interpretation of the prophecies in tedium.

The *Principia* marked a break in Newton's theological activities. There are a few theological manuscripts from the two following decades, but not many. Sometime during the first decade of the eighteenth century he returned to theology again, and he devoted massive amounts of time to it for the rest of his life. The effort, however, was concerned entirely with reshuffling old ideas to which he added nothing new. Like his career in science, his period of creativity in theology belonged entirely to his younger age.

A survey of the "other" Newton's activity in theology raises a number of questions that I want to address in conclusion. No one disputes the assertion that Newton was a scientist of major, indeed monumental, importance. What influence did his theology have on his science? I need to say that it is unclear to me that we can speak validly of any influence specifically of his theology, not his religion. The influence of his religion on his science is, I believe, universally admitted, and I do not challenge that conclusion. His theology, by which I mean explicitly his Arianism and the associated interpretation of the prophecies, is another matter. Perhaps we can find echoes of the Arian God in the Pantocrator of the "General Scholium," but this leaves us still on such a high level of generality that it tells us very little. If we want to descend to the details of Newton's science, as found in the *Principia* and the *Opticks*, I am unable to trace any line of influence that has substance.

Rather, I prefer to trace the influence in the other direction. Theology was the activity with the historically established role in European civilization, a role beginning to be challenged, for the first time in more than a millennium, by a newly rising enterprise, modern science. It appears to me that we are far more apt to find the lines of influence running in this direction.

During the last couple of decades, there have been a series of articles and at least one doctoral dissertation of which I know that have stressed the traditional elements in Newton's religion (Austin 1970; Trengrove 1966;

Brooks 1976). All of them were written before the Yahuda manuscripts were available, and consequently their authors called on a body of evidence different from the one I have utilized here. I like to think that the authors would have drawn the same conclusions from the Yahuda manuscripts that I have drawn, although, of course, I cannot presume to speak for them. In any case, all of the works in question stress the role of the Bible in Newton's religious thought. There is no doubt that Newton considered the Bible to be the word of God. Nevertheless, as I have indicated, what Newton had in mind when he said "Bible" was far from identical to what that word had meant to the previous Christian tradition. Of necessity, given the evidence on which they drew, all of these discussions of Newton's religion ignore what the Yahuda Papers, in my opinion, establish as the central fact, his Arianism. When I recall the role of Arianism in early Christianity and the role of its offspring, Unitarianism, in the modern world, I find it impossible to ignore the influence of science on his religion. What I have in mind when I say "science" is not what I understand as Newtonian science. To me, the concept of Newtonian science is associated indissolubly with the *Principia* and involves the critical notion of attractions. Newton assumed his characteristic theological position, however, before he had begun so much as to dream of the *Principia*. Therefore, when I speak of the influence of science on his religion, I am thinking of more basic stands associated with the scientific revolution, especially a new criterion of truth and a new locus on intellectual authority — all that Basil Willey had in mind when he spoke of "the touch of cold philosophy." The seventeenth century's touch of cold philosophy had many sources, but none of them appears to me more crucial than the rise of modern science. Like Boyle, Newton was aware that the ground under Christianity was shifting. The central thrust of his lifelong religious quest was the effort to save Christianity by purging it of irrationalities. If this interpretation is correct, it suggests the extent to which the "other" Newton would have appeared "other" to his own age. I am convinced that in Newton we have one of the earliest manifestations of what would be the most prominent response to Christianity during the Enlightenment. It is worth remembering in this respect that he was not only a seventeenth-century scientist; he was also an eighteenth-century theologian, still alive and writing theology when Voltaire visited London. But he was "other" to the Age of Reason as well, for Newton wrote, not from hostility to Christianity, but from a profound desire to purge it in order to preserve it. He was a man caught between two ages, one not yet ready to die, with which his affections remained, another struggling to be born, into which his intellect had largely entered.

The story of Newton and Christianity constitutes, in my perception of things, one chapter in the central drama of European civilization, the conversion of an originally Christian civilization into a scientific one. Newton was by no means alone; many others were wrestling with the same problems. Since Newton made every effort to keep his theological views private, his opinions did not enter prominently into the main channel of religious thought; insofar as they did enter, his endeavor to save Christianity by purging it only contributed to the ultimate change, which went far beyond anything he would have welcomed. How little even the greatest of us understand the consequences of our own actions! Anyone who has read the books I have written will know that I greatly admire the achievement that modern science represents. Nevertheless, I do not revel in the decline of Christianity. I am myself a practicing Protestant Christian and, as I look about me at the chaos of contemporary civilization, especially in its American manifestation, it is far from evident to me that the change has all been gain. Nevertheless, whether I or anyone else likes it or not, it is fact; it did happen — the greatest alteration that European civilization has undergone. It is not the least product of the scientific revolution, or of Isaac Newton.

Notes

1. Spence attributed this anecdote to Andrew Michael Ramsay (Spence 1820, 54).
2. Quoted from the Boyle Papers in the library of the Royal Society in Westfall 1958, 26.
3. After the auction of the papers owned by the Portsmouth family in 1936, the large majority of theological manuscripts passed into private possession and were ultimately willed to the Jewish National and University Library in Jerusalem, where they became available less than two decades ago.
4. Although there are a few dates in the theological manuscripts, one has to rely primarily upon the evidence of Newton's hand in placing them chronologically. In my article (Westfall 1982, 129–43), I included a checklist of the theological manuscripts with my considered opinion as to their dates. I shall draw upon that scheme of dating in the present account without attempting to repeat it all here.
5. This manuscript contains a version, which appears to date from the 1690s, of one chapter of the *Origines*.
6. The manuscript of that version was published after his death under the title *The System of the World*. It appears at the end of the Cajori edition of the *Principia*, with the passage to which I refer on p. 549.

References

Printed works of Newton

Opticks, based on 4th edition, New York, 1952.

The Correspondence of Isaac Newton, ed. H. W. Turnbull et al., 7 vols., Cambridge, 1959–1977.

Manuscript works of Newton

Babson Collection, Babson College, Babson Park, Mass.
Keynes Collection, King's College, Cambridge.
Paradoxical Questions Concerning the Morals & Actions of Athanasius and his Followers, Clarke Library MS, Los Angeles.
Yahuda Papers, Jewish National and University Library, Jerusalem.

Other references

Austin, W. H., 1970. "Isaac Newton on Science and Religion," *Journal of the History of Ideas* 31:521–40.
Brooks, R. S., 1976. "The Relationships between Natural Philosophy, Natural Theology and Revealed Religion in the Thought of Newton and their Historiographic Relevance,' Ph.D. dissertation, Northwestern University, Evanston, Ill.
Spence, Joseph, 1820. *Anecdotes, Observations, and Characters, of Books and Men*, ed. Samuel Singer. London.
Trengrove, L., 1966. "Newton's Theological Views," *Annals of Science* 22:277–94.
Westfall, R. S., 1982. "Newton's Theological Manuscripts," in *Contemporary Newtonian Research*, ed. Z. Bechler. Dordrecht: D. Reidel Publishing Company.

Newton and the Origins of Fundamentalism

RICHARD H. POPKIN

When I told an eminent American philosopher that I was working on Newton's religious writings, he first expressed the view that it was scandalous that the vast number of Newton manuscripts on the subject had not been published, and then asked me, "Frankly, are these writings of interest because they are by Newton, or are they important in their own right?"[1] To answer this, one has to consider what they represented in their own time, and what influence, if any, they had on theology then and later on.

Those who brush aside any interest in Newton as a commentator on Scripture, especially on the prophetic books of Daniel and Revelation, and as a defender of scriptural chronology, usually treat these concerns as a personal aberration of Newton, indicating that he was partially crazy, prematurely senile, or the like. However such scholars as Frank Manuel, Richard S. Westfall, J. E. McGuire, B. J. T. Dobbs, Charles Webster, James Force and others, who see Newton in the mainstream of English Bible interpreters as Thomas Brightman, Joseph Mede, Henry More and Ralph Cudworth, see that he was developing certain themes that were of paramount importance to seventeenth-century intellectuals, that he was answering certain problems that had arisen, and that he was breaking new ground in some of his interpretations.[2]

Newton was part of a tradition of seventeenth-century intellectuals who saw the increase of knowledge of nature and the deciphering of the mysteries of scriptural prophecies as a unified endeavor to understand God's creation and His Providence. As the end of the history of man and nature approached, God's plan in nature and in providential history would become ever clearer,

E. Ullmann-Margalit (ed.) The Scientific Enterprise, 241–259.

as prophesied in Daniel 12.[3] To appreciate Newton in his time as a Bible interpreter and theologian, one has to compare him with Joseph Mede, the great decoder of the book of Revelation for seventeenth-century English millenarians like More, Cudworth, John Milton and Isaac Barrow.[4] Newton was influenced by and worked with Henry More in developing his interpretation of the prophecies.[5] Newton's one-time disciple, William Whiston, probably voiced some of Newton's private views in public. At least, Whiston said that he did (see Force 1985; Whiston 1749), and some of the Yahuda papers (the unpublished writings of Newton in the Jewish National and University Library at Jerusalem) indicate there is some truth to this.[6] One should also evaluate Whiston's contribution to seventeenth-century Millenarian theology. In this context, I believe, Newton stands out as one of the giants of his time. However, I will put this aside here, in order to concentrate on the long range impact of some of his views on the development of what is now called Fundamentalism.

Frank Manuel, the only one to write a book entitled *The Religion of Isaac Newton,* and the only person to publish any of the Yahuda papers in Jerusalem, excused himself as a historian, rather than a theologian, for writing on Newton's religious views. His excuse was that "Newton's printed religious views have exerted no profound influence on mankind" and were only cited by religious apologists in the eighteenth and nineteenth centuries to demonstrate the compatibility of science and faith. An interest in Newton's religious views, Manuel insisted, can hardly be justified by their value for the propagation of faith (Manuel 1974, 1–2). On the contrary, I shall try to show that Newton's views, or part of them, had a most serious and important impact on one aspect of what is now called Fundamentalism.

Little attention has been given to Newton's influence as a theologian in the latter part of the eighteenth and early nineteenth centuries. Two of his works on religious topics had been published, *The Chronology of Ancient Kingdoms Amended* (1728) and the posthumous *Observations on Daniel and Revelation* (1733). A third work, two letters written to John Locke, and sent to Jean LeClerc, against the textual evidence for the doctrine of the Trinity, were published in 1743. All three items were republished in Bishop Horsley's edition of Newton's *Opera* (vol. 5), with the latter item corrected and amended from the original holograph copy. During the period after Newton's death, his religious views were reported on, commented on, and debated by Whiston and a host of other minor theologians. They apparently influenced important figures like Swedenborg, J.G. Hamann, David Hartley, Joseph Priestly and Samuel Taylor Coleridge (cf. Manuel 1974, 1–2; Popkin 1984). These views became central for an emerging resurgence of Millenarianism at the time of the American and French revolutions. The formulation of biblical interpretations by many of those who perceived these revolutions as events prophesied in Daniel and Revelation has been recognized as an early statement of what is now called Fundamentalism. Newton, as we shall see, appears as one of the theological heroes, along with Mede, More, Whiston and Bishop Thomas Newton, who foresaw the imminent beginning of the Millennium in the first half of the nineteenth century in terms of the fulfillment of premillennial prophecies. Some of these theologians founded the organizations which developed into the structure of modern Fundamentalism, and offered justifications for their interpretations, based in part on Newton, that became part of the Fundamentalist thesis.

First, a few words about Fundamentalism and its history. The term was given its present meaning around 1920, when some American Protestant theologians set down what they considered the fundamentals that a true and believing Christian had to accept. The term does not appear in older editions of the Oxford English Dictionary, and is a twentieth-century designation of a kind of conservative, Millenarian view, espoused in the face of nineteenth-century biblical criticism and science (especially Darwinism). One of the chief historians of the movement, the late E. R. Sandeen, traced the basic content of Fundamentalism back to certain British Millenarians, who interpreted the events of the French Revolution and the Napoleonic era as the last important premillennial events (Sandeen 1968). The historian of the Seventh Day Adventists, Leroy E. Froom, in his four-volume *Prophetic Faith of Our Fathers* (1950–54), covers the whole history of Millenarianism back to early Christian times, and stresses in his third volume the developing theology in Colonial and Revolutionary America, and in Britain from 1789 onward, as the background to the rise of his own church and that of nineteenth century fundamentalist currents. Froom, of course, has a chapter on Newton as a seventeenth-century Millenarian (Froom 1950–54, 2:658–69). I have uncovered other significant American and English writers, not discussed by Sandeen or Froom, who betray their debt to Newton even more clearly, and have considered their contribution to the formulation of Fundamentalism. For the purposes of this study I will confine myself to two major Fundamentalist themes — the accuracy and inerrancy of Scripture, and Scripture as the coded statement of the future course of history.

In the aftermath of the orgy of millennial writings during the Puritan Revolution, a more restrained presentation, developed by Napier, Brightman, Alsted, and especially Joseph Mede, produced a calculus for dating the imminent onset of the Millennium. Dates settled on 1648, then 1655–56, then 1666 (cf.Popkin 1980; Hill 1988; Katz 1982). As each date passed, theorists had to restudy the texts of Daniel and Revelation. Some continued to propose new dates, seeing the Glorious Revolution, various earthquakes, etc., as signs that the end was near. Others, especially Henry More and his student and friend, Sir Isaac Newton, offered a more restrained view. More, having fended off the Puritans, the Quakers and other enthusiasts, set forth an interpretation of Daniel and Revelation that put the crucial millennial events somewhere in the indefinite future. Newton, in his posthumously published *Observations on Daniel and Revelation*, was both more precise and vaguer. He showed in inordinate detail how post-biblical history through the Middle Ages was forecast or prophesied in Daniel's statements about the four monarchies, and in the opening of the seals in Revelation. He did not deal, in the published work, with Reformation and seventeenth-century events. Instead, he explained that prophecies could only be recognized *post factum*, by studying events and seeing if they were part of the premillennial scenario:

> The folly of interpreters hath been, to foretell times and things by prophecy, as if God designed to make them prophets. By this rashness they have not only exposed themselves, but brought prophecy also into contempt (*Observations*, 251.)

However, God's purpose through prophecies was to make men realize what was going on.

> After they (the prophecies) were fulfilled, they might be interpreted by the event, and his own Providence, not the interpreters, be then manifested thereby to the world. For the event of things, predicted many ages before, will then be a convincing argument that the world is governed by Providence. (Ibid., 251–52.)

Thus, many prophecies have been fulfilled. Others will be, concerning the Second Coming of Christ and the millennial kingdom. But we can only understand progressively, as God allows us, how these events are coming about. The time has not yet come to understand the prophecies perfectly. But so much has happened that fits the prophetic predictions, that we can expect that

> the signal revolutions predicted by all the holy prophets, will at once both turn men's eyes upon considering the predictions, and plainly interpret them. Till then we must content ourselves with interpreting what hath been already fulfilled. (Ibid., 252–53.)

Newton acknowledged that most seventeenth-century interpreters of Daniel and Revelation had made worthwhile discoveries. This indicates, he said, "that God is about opening these mysteries" (ibid., 253). But Newton did not state in public when this might happen.

In public, Newton promoted a careful examination of past prophetic fulfillments, and the expectation that the remaining ones would occur soon. The Boyle lectures of Whiston and Samuel Clarke offered a kind of inductive Millenarianism, by pointing out how many Old Testament prophecies had been fulfilled from the time of St. Paul until the present (Whiston 1708; Clarke 1705). Hence, reasonable people should expect the fulfillment of the great remaining prophecies about the conversion of the Jews, the reestablishment of the Jewish Kingdom in Jerusalem, the Second Coming, the Millennium and the end of the world. Reasonable people should inspect events to see if they were part of this fulfillment.

Whiston became the butt of ridicule for seeing providential signs in every natural catastrophe (cf. Force 1985, ch.1). Others, such as David Hartley, the founder of associationist psychology, and Bishop Thomas Newton, put forth Isaac Newton's view without making exact claims as to when the great events would occur (Hartley 1749, pt.II; T. Newton 1766).

In the face of English deist and French Enlightenment critiques of the truth and accuracy of Scripture, some religious writers in America and England fought back because they were convinced that the events dominating their worlds were the events prophesied in Daniel and Revelation as the events that would immediately precede the Millennium. Fairly important figures, such as Joseph Priestley, argued that the end-of-the-world scenario of the prophecies was evidently in process in the latter part of the eighteenth century.[7] Bishop Thomas Newton had given the Boyle Lectures in the 1750s on prophecies and had indicated that the final ones were about to be fulfilled.[8]

This kind of view was taken up by two interesting figures who interpreted the American Revolution as the actual fulfillment of these prophecies and a group of figures similarly interpreting the French Revolution and the Napoleonic period. The authors I will consider begin the theological current that blossomed in the twentieth century as Fundamentalism, and they openly acknowledge an important debt to Isaac Newton's religious writings.

First, I will outline the views offered and the references to Newton, subsequently trying to assess what it was in Newton's views that was important to these protofundamentalist theologians and how this has developed. These writers usually cite Newton, Joseph Mede, Henry More and Bishop Newton. There is a part of Isaac Newton's published religious writings which was especially significant for these thinkers, and which was different, at least in emphasis, from what they found in the other heroes of millennial thought, namely, Newton's view that one could discern *post factum* where one was in the Daniel-Apocalypse scenario.

A leading figure in the American Revolution, Elias Boudinot (1740–1821), a lawyer in New Jersey of Huguenot descent, who became President of the Continental Congress that governed the American Colonies during the Revolution, and who signed the peace treaty with England, was, perhaps, the most important Millenarian layman in the country. He had been baptized by the great evangelist revivalist Whitfield. He actively participated in the Revolution and was the patron of Alexander Hamilton, who had lived with him as a youth. Boudinot was elected to the first United States Congress and then became the first Director of the Mint, the same position that his Millenarian predecessor Isaac Newton had held in England. Boudinot resigned from the government in 1805 when Jefferson became President, explaining that he could not work for a deist government, and that he had too much work to do before the Millennium. He founded two of the organizations that were the forerunners of Fundamentalism in the United States: the American Bible Society and the Society for Ameliorating the Condition of the Jews.[9] He published three important works, *The Age of Revelation* (1801), *The Second Advent* (1815), and *A Star in the West* (1816). The first was an answer to Tom Paine's *Age of Reason*, in which Boudinot sought to show that it was obvious to reasonable people that the world had developed from Mosaic history to the present, and that there was no good reason to doubt the accuracy of the scriptural account or to doubt that present history was the fulfillment of that account. The book ends saying, "the evidence in favour of the great truths of revelation are daily increasing, and will so continue till the second coming of the Lord Jesus, as he has promised" (Boudinot 1801, 329).

The Second Advent, or the Coming of the Messiah in Glory shown to be a Scripture Doctrine, and Taught by Divine Revelation, from the Beginning of the World, by an American Layman shows Boudinot following Newton's advice about interpreting events as fulfillments of prophecies. In the preface, Boudinot says that in 1790 events of the time led him to reexamine scripture. He studied the prophecies. Daily events alarmed him as "he thought he saw signs of the latter day, foretold in sacred record, thickening upon him" (Boudinot 1815, preface). He kept a diary of events in Europe from 1790 onward.

> The many instances of exact conformity with the words and spirit of Scripture, convinced the author that the wonderful transactions daily passing in the kingdoms of Europe, were an exact fulfillment of the

> predictions of the Sacred record. That the Antichrist foretold, as coming on the earth after the Man of Sin, had literally appeared in the new government of France, having Napoleon Buonaparte for her head, can scarcely be denied by any observing mind, who has become acquainted with the late history of that nation since the year 1790, and compared it with the language of holy writ. (Ibid.)

In developing his prophetic reading of contemporary history, Boudinot appealed many times to Isaac Newton's opinions and to Newton's view about the fulfillment of prophecies, as well as to the views of Mede and Bishop Newton. In *A Star in the West* he portrayed what he saw as America's role in providential history: to abolish slavery, liberate and convert the Jews, and reveal the Indians as the Lost Tribes — all in preparation for the great events to come. Boudinot's activities in the American Bible Society and the Society for Ameliorating the Condition of the Jews sought to further this end. He was also an advocate of female equality for millenarian reasons.[10]

A fellow abolitionist, Charles Crawford, saw the providential character of events sooner, and considered Newton as an important authority in interpreting contemporary events. Crawford was born in Antigua in 1752 and attended Queen's College, Cambridge. He was in America around 1783, urging people "to forbear to reproach the Jew, to free the Negro, to tame and incorporate the Indian, and to invite them all to Christianity".[11] In 1784 he published an important abolitionist work, *Observations on Negro-Slavery*, in which he argued against David Hume for black equality.

Then he foresaw the conversion of the Jews "before a very great length to time," as well as the recognition that the Indians were descended from the Lost Tribes. The last point would make people treat the Indians decently.[12] In the course of this humane discourse Crawford mentioned that Isaac Newton and Samuel Clarke were the best Bible commentators. He took issue with them about when the Papacy would fall, placing the event around 1848, "notwithstanding the high reverence which I bear for the learning and talents of Sir Isaac Newton and Doctor Clarke" (Crawford 1801, 111).

His first serious reading of the providential signs of the time appeared in his *Observations upon the Downfall of the Papal Power and Consequent End* (1785), dedicated to the Emperor Joseph, because of the latter's act of toleration to the Jews, who were about to be restored. Crawford began his exposition by citing Newton on the downfall of empires. A few years later he took cognizance of the millennial implications of the French Revolution in his *Observations on the Revolution in France* (1793). Mede's schema, in his *Clavis Apocalyptica*, was used as the basis for interpreting current events.

Newton's *Observations* were cited concerning the potential downfall of Papal power, which Crawford now said would be followed by the conversion of the Jews and their restoration to the land of their forefathers. In this work, Crawford, unlike the others I will discuss, cited Newton's letter to Locke about the doctrine of the Trinity, and apparently approved of Newton's anti-Trinitarianism.[13]

Crawford's use of Newton's ideas appears in two other works. In 1800 he published *An Essay upon the Eleventh Chapter of the Revelation*. Here he made Mede the great commentator, deeming Newton not quite as diligent. Newton's efforts, he reported, were found among the sweepings of his library. There is no evidence that Crawford knew how much of Newton's theological writings had survived in the papers left at his death. Mede, Crawford knew, had worked twenty years examining the text of Revelation; he supposed that Newton's efforts were more casual.

Crawford realized before the French Revolution that events taking place in America and Europe were following the course set out in Daniel and Revelation. He was therefore active in premillennial causes, abolitionism, emancipation of the Jews and equal rights for Indians. On the other hand, like Boudinot, he was appalled by the irreligious efforts of people like Tom Paine and other French and American deists and atheists (though this apostasy from the true faith was part of the prophesied scenario). Crawford left the United States in 1802, to spend the rest of his life in England.[14] There he published *Three Letters to the Hebrew Nation* (1817), in which he foresaw the conversion of the Jews and their restoration to Palestine by 1900. He was trying to prepare them for this transformation. In presenting his evidence that these events were imminent, he cites much material from Mede, Newton and Clarke, citing them as the most judicious modern commentators. Though finding fault with them on detail, he held that they had presented the format for interpreting the events that he, Crawford, was living through, in which Napoleon was being used by false prophets to lead the Jews astray. Crawford, working with the English Society for Converting the Jews, saw Newton as an aid in protecting the Jews from false doctrines (Crawford 1817, 78–79). He specifically recommended that they read Newton's anti-Trinitarian letters to Locke (ibid., 113n). In terms of practical affairs, Crawford hoped that Czar Alexander I would help bring about the conversion of the Jews, by helping the Bible Society to distribute Hebrew copies of the New Testament and by ameliorating the condition of the Jews of eastern Europe in the post-Napoleonic era (ibid., Appendix, 117–120).

Crawford, like Boudinot, became convinced, from American historical experience and the events of the French Revolution, that mankind was now living through the events forecasted by Daniel and John. He used the best commentators he could to understand what was happening and what was to be done. Newton was a major influence for his *post factum* approach to interpretation, for his actual interpretations and for his doctrinal purity, that is, his anti-Trinitarianism, which was only appreciated by Unitarians of the time and ignored by the other protofundamentalists.

The strongest drive to interpret events prophetically occurred in England, as a response to the French Revolution and Napoleon. The English and Scottish societies for converting the Jews and for propagating the Gospel developed a more coherent view and practice, which led ultimately to the Fundamentalism of the late nineteenth century. The English efforts often dovetailed with American ones, and Millenarian authors like Hannah Adams of Boston were published by both American and British societies.

The number of authors one could examine among the English are legion. To illustrate Newton's influence, I will select only three important figures, George Stanley Faber (1773–1854), William Cuninghame (1776–1849), and James Hatley Frere (1779–1886). All three were actually involved in the London Society for Promoting Christianity amongst the Jews and in formulating the evangelical position that was drawn up at the Albury Park conference, which became the central protofundamentalist position in the first half of the nineteenth century. Moreover, all three were mature by the end of the Napoleonic era and saw the history in which they had grown up as a crucial part of providential history that would soon (by mid-century, they thought) lead to the actual Millennium. They all became involved in trying to calculate the exact date of the end of days (cf. Sandeen 1968; Froom 1950–54 3: chaps. xviii, xx).

Faber was an active church leader and writer. He became prebendary of Salisbury Cathedral, and in the course of his long life wrote over forty books. One of his central themes was that modern history is apocalyptic history. One of his most famous works, which went through five editions, was *Dissertation on the Prophecies that have been fulfilled, are now fulfilling, or will hereafter be fulfilled, relative to the great period of 1200 years, that Papal and Mahometan Apostacies, and tyrranical reign of Antichrist, or the Infidel Power, and the Restoration of the Jews*, in two volumes (first published in 1806, fifth edition 1814).[15] In this work and in *Prophecies relating to Judah and Israel* (1808), he sought to analyze Napoleon's career and activities in terms of the accounts in Daniel and Revelation. He used many commentators, adopting Newton's claim that a year in the Jewish calender equals a solar year (Faber 1811, XV, 49–65; see Froom 1950–54, 3:344). He contended that exact calculation of when the prophesied events would occur was not possible, but that very plausible conjectures could be made. As history proceeded after Napoleon, Faber continued to interpret it in providential terms. He lived long enough to see Louis Napoleon crowned Emperor of France, and his last work, published in four editions in 1853, was *The Revival of the French Emperorship anticipated from the necessity of Prophecy.*[16]

Cuninghame was a polemicist for the London Society, arguing chiefly with Jewish opponents. He published copiously in the proceedings of the Society and in their magazine, *The Jewish Expositor*, against David Levi,[17] the leading English defender of Judaism in the late eighteenth century, and against Rabbi Joseph Crool, a contemporary.[18] Cuninghame perceived present events as those foretold in Daniel and Revelation. In his *Political Destiny of the Earth as Revealed in the Bible* (1833), he said he was dealing with heavenly politics "of that kingdom which is about to be revealed."[19] Before setting out his interpretation of events from the overthrow of Louis XVI to the present, he outlined Daniel's prophecies and cited Newton as his main authority on interpreting the text (cf. Froom 3:368, 427). Newton, Mede, More, Brightman, Whiston and Bishop Newton were his main sources for his own interpretation. Based on his reading of events, he concluded, "That the present moral and political condition of the world is altogether without parallel in the past history of mankind will be denied by no accurate observer" (Cuninghame 1842, 26–27).

The last English theologian I shall consider, James Hatley Frere, was the theoretician of the evangelical movement that centered around the London Society and the Albury Park Conference. He greatly influenced Joseph Wolff, Lord Drummond and Edward Irving, and he in turn betrays the greatest influence of Newton on his ideas. One of his most important works, *A Combined View of the Prophecies of Daniel, Ezra and St. John; showing that all the Prophetic Writings are formed upon one Plan*[20] deals in great detail with Napoleon as symbolized in the prophecies, and with the Holy Roman Empire (as does Irving in his *Babylon and Infidelity Foredoomed of God. A Discourse on the Prophecies of Daniel and the Apocalypse which relate to these latter times and until the Second Advent* [1826], dedicated "To my beloved friend, and brother in Christ, Hatley Frere"). Frere cited some of the usual list of commentators — Mede, Bishop Newton and Whiston. However, he repeatedly emphasized the special role of Isaac Newton in his view. He wrote, "I am gratified to find, that I have such high authority as that of Sir Isaac Newton" for his interpretation of one image in Daniel (Frere 1826, 151). He expressed his happiness that Sir Isaac and Bishop Newton agreed with him on the role of the Exarchate of Ravenna (ibid., 166). And there were other points that happily he and Newton agreed on. It is interesting that Frere never said that Newton, the eminent scientist, agreed with him, but rather that Newton, the eminent commentator on Scripture, agreed with him. Frere's disciple, Irving, never mentioned Newton. Irving cited very few authorities; as an ardent Trinitarian, he may have purposely omitted Newton.

The writers discussed above, plus their allies and codiscussants, developed a great expectation that the climax of human history would occur in the 1840s. From accepting Newton's *post factum* method, they moved on to calculate from the American and French revolutions, from Napoleon's Egyptian campaign, his capture of Rome and the Pope, and his fall, when the last premillennial events would occur. "The Great Disappointment"[21] of the 1840s did not lead to the abandonment of prophetic interpretation, but rather to its incorporation in new Fundamentalist sects, such as the Seventh Day Adventists and Jehovah's Witnesses, and in the formulation of early Fundamentalist platforms. From one of these, the Niagara creed of 1878 (cited in Sandeen 1968, 273), one can see both the positive legacy that is in part drawn from Newton, and another aspect that diverges from Newton's views. The last article of this creed, which was taken over by most later Fundamentalist groups, states:

> We believe that the world will not be converted during the present dispensation, but is fast ripening for judgment, while there will be a fearful apostacy in the professing Christian body; and hence that the Lord Jesus will come in person to introduce the millennial age, when Israel shall be restored to their own land and the earth shall be full of the knowledge of the Lord, and that this personal and premillennial advent is the blessed hope set before us in the Gospel for which we should be constantly looking.

This statement does not attempt to set any particular time for the occurrence of these great events, but stresses the need for "constantly looking." When the events are observed, then, and only then, will one be able to interpret them as part of providential history.

As mentioned previously, Leroy Froom, the Seventh Day Adventist historian, devoted a chapter of his *The Prophetic Faith of Our Fathers* to Newton. Froom stresses that Newton employed the same rigor in finding the facts about prophecy, and applied the same strictly logical method of deduction, that he used in mathematics and physics.[22] Newton saw Daniel, Froom says, as the key to all other prophecies, and saw that Christianity rests on them and can only be understood in terms of them. Newton, we are told, was superior to Joseph Mede in interpreting the prophecies (Froom 1950–54, 2:660–664). But, Froom stresses by quoting Newton,

> The folly of Interpreters (other than Newton) has been, to foretell times and things by this prophecy, as if God designed to make them Prophets. By this rashness they have not only exposed themselves, but brought the Prophecy also into contempt. (Ibid., 665.)

And, as cited earlier, for Newton, God's purpose was to enable men not to foreknow things, but to recognize that certain things, when they occurred, manifested God's Providence (ibid., 665–66). The passage in Newton liberated the interpreters of prophecy from having to be right about predictions of when things would happen. Instead, as the Niagara creed stated, they had to look at events constantly to see if the prophecies had been fulfilled. Newton and the early Fundamentalists were sure that Daniel and Revelation contained predictions of providential history and signs of it to be sought. Newton, as an acute commentator, contributed to the detailed exposition of the prophecies. But he also contributed a *post factum* kind of inquiry. During the early nineteenth century, various American and English interpreters thought they could identify the signs in the revolutions and in Napoleon's career, and could then calculate when other predicted events would occur. The debacle of the predictions of the 1840s led to a reconsideration. The Niagara creed incorporates Newton's caution. They would not predict; they would look. They knew what they were looking for, namely, the Second Advent and the Restoration of Israel. After attempts to find providential significance in World Wars I and II and other early-twentieth-century events, the Restoration of Israel, "when Israel shall be restored to their own land," did indeed occur, when the State of Israel was founded. For many Fundamentalists, it is now obvious that the end of providential history is at hand. Countless volumes are pouring out about present events in the Middle East and how they relate to the prophecies in Daniel and Revelation. The alliance of Zionists and Fundamentalists has been born from this, and we may be in state for another "Great Disappointment."[23] If so, Newton has still provided a way of accounting for human error and for persevering in further historical researches.

Another well-known side of recent Fundamentalism strays from Newton's views, namely the stress on the "inerrancy of Scripture." In the published *Observations* Newton both accepted Scripture as true and agreed that we might have erroneous texts. He was aware of much seventeenth-century biblical research; he read Richard Simon's *Critical Histories of the Old and New Testaments*, through which he would have learned of Spinoza's criticisms of the biblical text had he not encountered Spinoza's views directly or through Henry More's criticisms of them.[24] Newton's *Chronology of Ancient*

Kingdoms Amended certainly shows that he was aware of the arguments about whether there were pre-Adamite cultures, predating the Bible. The *Chronology* is an attempt to vindicate biblical chronology in the face of knowledge about other ancient cultures, and to show that there is no reason to doubt the accuracy of the Bible as the history of all mankind.[25]

At the beginning of the *Observations on Daniel and Revelation*, Newton presented his account of the history of the various books of the Bible. He did not claim that the Pentateuch, as we know it, was directly revealed by God to Moses. Instead, he gave a complicated account of how the text was compiled over centuries. He dated the text as we know it after the Roman captivity, from talmudic times. As a result,

> the ancienter lections, except what can be discovered by the errors of the transcribers, before the edition was made, had crept into the text, are now scarce to be corrected. (*Observations*, 11–12.)

Hence, Newton, in his published writings, was willing to accept the first germs of biblical criticism that had appeared in his time and to agree that the existing text might involve errors. Of course, his letters to Locke are a model of biblical criticism, showing the corruption of the texts that deal with the Doctrine of the Trinity. He told Locke "it's manifest that ye scriptures have been much corrupted in ye first ages & chiefly in the fourth century, in the times of the Arian Controversy."[26] Newton did not accept the claim of the Westminster Confession that God had preserved the text through history. The Westminster Confession declared that the authority of Scripture rested on the fact that God was the author. The text in Hebrew and Greek, "being immediately inspired by God and by his singular care and Providence kept pure in all Ages are therefore Authentical" (*Westminster Confession*, 6). Newton was willing, instead, to accept the best findings of historical scholarship about the human alterations that may have been made in the text in the course of Jewish and early Christian history. To this extent, he was a liberal in the context of the arguments of his time, veering away from the literalism of the Calvinists and the attempt to avoid any historical investigation of the Bible. In terms of ninteenth- and twentieth-century Fundamentalism, Newton could hardly agree with such formulations as that of the Niagara creed, Article I:

> We believe that all Scripture is given by inspiration of God, and that this Divine inspiration extends equally and fully to all parts of the Biblical writings, historical, poetical, doctrinal and prophetical, and to the smallest word, and inflection of a word, provided such word is found in the original manuscripts. (Cited in Sandeen 1963, 273.)

For Newton there were no original manuscripts on which to base decisions as to the actual text.

Of course, Newton was not fighting the same battles as the modern Fundamentalists. He assumed he had satisfactorily established the accuracy of scriptural history and the role of prophecy as the core of providential history. Texts he deemed to be human copies, subject to human failings, were to be adjudged by the best scholarly standards. The modern Fundamentalists, after two centuries of biblical criticism, have entrenched themselves in an ahistorical position, resisting all challenges from historical scholarship and science.

To conclude, I hope my evidence has shown that Newton, through his three published theological writings, played a significant role in religious history. His interpretations of the prophetic writings and his *post factum* theory of interpretations were used by the late eighteenth- and early nineteenth-century American, British and Scotch theologians who established a basic part of the modern Fundamentalist view. Further investigation of a wider range of religious writers may make it possible to delineate more carefully the true boundaries of Newton's contribution. But I think one has to admit that he was read and used by those who formulated the reading of modern history, from the American and French Revolutions onward, as the beginning of the end of days. His method of avoiding disconfirmation by stressing *post factum* interpretation has provided religious Fundamentalists with a way to continue reading events in prophetic terms, regardless of the lack of success of such attempts in the past. The present resurgence of Fundamentalism since the founding of the State of Israel demonstrates how Newton's view can buttress the hopes end spirits of those who accept Daniel and Revelation as the blueprint of history. In these terms, Newton may have made as great a contribution to Bible interpretation as he did to the interpretation of Nature.

Notes

1. The question was raised by one of my teachers, Paul Weiss.
2. Cf. Manuel 1963; 1974; Westfall 1980; McGuire and Tamny 1983; Dobbs 1975; Webster 1975; Force 1985.
3. Daniel 12:5: "O Daniel, shut up the words and seal the book *even* to the time of the end: many shall run to and fro, and knowledge shall be increased."
4. Cf. Firth 1979; Froom 1950–54, vol.2; Tuveson 1972; Popkin 1986a, 1988; Webster 1975.
5. Newton and More had a serious disagreement in 1680 about interpreting the book of Revelation, and thereafter they seemed to have stopped working together. See More's letter to Dr. John Sharp, August 16, 1680, published in Nicolson 1930, 478–79. The

Bancroft Library at the University of California, Berkeley, has Newton's copy of one of More's works on Revelation with negative comments written in it. Christ's College, Cambridge, has a copy of More's last work on Revelation inscribed to Newton. Since it was in More's library, Newton may have refused the gift.

6. Yahuda Ms. Var. 1 consists of at least half of Newton's unpublished writings, principally on theological subjects. It contains many pieces that set forth Newton's millenarian and anti-Trinitarian writings more strongly than his published works. Professors Dobbs, Westfall and myself are heading up a project to publish all of Newton's alchemical and theological writings.
7. See Priestley 1787, 1789, 1794. Cf. Popkin 1984, 50–52.
8. He was appointed to give the Boyle Lectures in 1754. They were published first in three volumes in 1754–58.
9. On Elias Boudinot's career, see Boyd 1952 and Whittlesley 1981. Boyd (1952, 261) writes that Boudinot was "without a peer as the foremost Christian layman of the United States."
10. Cf. Boyd 1952, 255–61; Boudinot 1816. On 23 September 1783, Boudinot gave a speech to the Society of Cincinnati, advocating the rights of women. The speech is reprinted in Boudinot 1971, 354–61.
11. Crawford is an interesting figure, who has hardly been studied. He was a radical abolitionist, anti-Trinitarian, philo-Semite and millenarian. Because his career was partly in revolutionary America and partly in England thereafter, he seems to have been missed by most scholars of American or British history. He is briefly treated in Jordan 1968 and in Bloch 1985.
12. Crawford 1784; 1801. In the latter, p. 23, he said that the prophecies about the recall of the Jews "should induce the whole people of America to treat the Indians with as much levity and forebearance as possible."
13. The other prophetic interpreters of the American and French revolutions seem to have ignored the overt evidence that Newton, Whiston and Clarke were all anti-Trinitarian Arians.
14. Crawford's role in the work of English missionary societies is discussed in a doctoral dissertation that Sarah Kachav is preparing at Oxford. She has found much new material about Crawford's career after he left America.
15. The work was first published in London in 1806. Some revisions were made in each subsequent edition, relating to historical events that had just taken place.
16. In the same year he wrote *The Predicted Downfall of the Turkish Power. The Preparation for the Return of the Ten Tribes.*
17. On David Levi, see Popkin 1987 and "David Levi, Anglo Jewish Polemist," forthcoming.
18. Rabbi Joseph Crool was active in England in the middle of the first half of the nineteenth century. He argued against the views of the Conversionist London Society for Propagating Christianity amongst the Jews, and for a while gave lectures at Cambridge.
19. On Cuninghame, see Froom 1950–53, 3:364–81.
20. On James Hatley Frere, see Froom 1950–54 3:385–90. *The Combined View of the Prophecies of Daniel, Ezra and St. John* was published in London in 1814, 2nd. ed. in 1826, London.
21. On "The Great Disappointment" see Froom 1950–54, chap. 40; Sandeen 1968.
22. Froom 1950–54, 2:659–60. Pages 658–69 are devoted to Newton as an expositor of prophecies.
23. Such volumes have been appearing from 1948 to the present, written by Fundamentalist evangelical authors, interpreting the latest events in the Middle East and the Cold War in these terms. See, for instance, the widely read *The Late Great*

Planet Earth, by Hal Lindsey, 1970, which seems to have influenced Ronald Reagan, as I have tried to show in Popkin 1986b.

24. There has been some dispute about whether Newton read Spinoza's *Tractatus Theologico-Politicus*. No citation has been found. Cudworth, More and Isaac Barrow all read it, and Newton no doubt heard about Spinoza's views from them. In a forthcoming essay I deal with Newton as a Bible critic and compare his views about textual matters in Scripture with those of Spinoza, Richard Simon, and others.
25. On this, see Manuel 1963; Ross 1984.
26. Newton's letters to John Locke, beginning 4 November 1690, in *Correspondence*, 3:83–146.

References

I. *Works of Newton*

Chronology: The Chronology of Ancient Kingdoms Amended. London, 1728.

Correspondence: The Correspondence of Isaac Newton, ed. H.W. Turnbull. Cambridge, 1961.

Observations: Observations upon the Prophecies of Daniel and the Apocalypse of St. John. London, 1733.

Opera: Opera quae exstant omnia, Commentariis illustrabat S. Horsley. London, 1779–85.

Two Letters: Two Letters to ... M. Le Clerc ... the former containing a dissertation upon the reading of the Greek text, 1 John v.7; the latter upon that of 1 Timothy viii.16. London, 1754.

II. *Other authors and scholarly literature*

Bloch, R. H., 1950. *Visionary Republic. Millennial Theories in American Thought, 1756–1800.* Cambridge.

Boudinot, E., 1801. *The Age of Revelation, or the Age of Reason shown to be an Age of Infidelity.* Philadelphia.

—, 1815. *The Second Advent, or the Coming of the Messiah in Glory shown to be a Scripture Doctrine, and Taught by Divine Revelation from the Beginning of the World by an American Layman.* Trenton, N.J.

Boudinot, J. J., ed. 1971. *The Life, Public Services, Addresses and Letters of Elias Boudinot.* New York.

Boyd, G. A., 1952. *Elias Boudinot, Patriot and Statesman, 1742–1821.* Princeton.

Clarke, S., 1705. *A Demonstration of the Being and Attributes of God.* London.

Crawford, C., 1784. *Observations on Negro Slavery*. Philadelphia.
—, 1788. *Observations on the Downfall of the Papal Power and Consequent End*. Philadelphia.
—, 1793. *Observations upon the Revolution in France*. Boston and Philadelphia.
—, 1800. *An Essay upon the Eleventh Chapter of the Revelation, in which it is shown that the words "And in the same hour was there a great earthquake, and the tenth part of the city fell, and in the earthquake were slain of men seven thousand" relate to Jerusalem, and not to Rome or France*. Philadelphia.
—, 1801. *An Essay on the Propagation of the Gospel; in which These and Numerous Facts and Arguments Adduced to Prove that Many of the Indians in America are Descended from the Ten Tribes*. Philadelphia.
—, 1817. *Three Letters to the Hebrew Nation*. London.
Cuninghame, W., 1842. *The Political Destiny of the Earth as Revealed in the Bible*. London.
Dobbs, B. J. T., 1975. *The Foundations of Newton's Alchemy, or "The Hunting of the Green Lyon."* Cambridge.
Faber, G. S., 1808. *A General and Connected View of the Prophecies, Relative to the Conversion, Restoration, Union and Future Glory of the Houses of Judah and Israel; the Progress, and Final Overthrow, of the Antichristian Confederacy in the Land of Palestine; and the Ultimate General Diffusion of Christianity*. London.
—, 1853a. *The Revival of the French Emperorship Anticipated from the Necessity of Prophecy*. London.
—, 1853b. *The Predicted Downfall of the Turkish Power. The Preparation for the Return of the Ten Tribes*. London.
Firth, K. R., 1979. *The Apocalyptic Tradition in Reformation Britain 1530–1645*. Oxford.
Force, J. E., 1985. *William Whiston: Honest Newtonian*. Cambridge.
Frere, 1826. *The Combined View of the Prophecies of Daniel, Ezra and St. John*. London. (1st ed. London 1814.)
Froom, L., 1950–54. *The Prophetic Faith of our Fathers*, 4 vols. Washington, D.C.
Hartley, D., 1749. *Observations on Man. His Frame, his Duty and his Expectations*. London.
Hill, C., 1988. "Till the Conversion of the Jews," in Popkin 1988, 12–36.
Irving, H., 1826. *Babylon and Infidelity Foredoomed of God*. Glasgow.
Jordan, W., 1968. *White over Black*. Chapel Hill. N.C.
Katz, D. S., 1982. *Philo-Semitism and the Readmission of the Jews to England 1603–1655*. Oxford.

McGuire, J. E., and M. Tamny, 1983. *Certain Philosophical Questions: Newton's Trinity Notebook*. Cambridge.

Manuel, F., 1963. *Isaac Newton, Historian*. Cambridge.

—, 1974. *The Religion of Isaac Newton*. Oxford.

Newton, T., 1766. *Dissertations on the Prophecies which have Remarkably been Fulfilled, and at this Time are Fulfilling in the World*, 3rd ed. London.

Nicolson, M. H., 1930. *Conway Letters: The Correspondence of Anne, Viscountess Conway, Henry More and their Friends 1642–1684*. New Haven.

Popkin, R. H., 1980. "Jewish Messianism and Christian Millenarianism," in *Culture and Politics from Puritanism to the Enlightenment*, ed. P. Zagorin. Berkeley and Los Angeles, 67–90.

—, 1984. "Divine Causality: Newton, the Newtonians and Hume," in *Greene Centennial Studies, Essays Presented to Donald Greene in the Centennial Year of the University of Southern California*, ed. P.J. Korshin and R.R. Allen. Charlottesville, 40–56.

—, 1986a. "The Third Force in Seventeenth-Century Thought: Skepticism, Science and Millenarianism," in *The Prism of Science*, ed. E. Ullmann-Margalit. Dordrecht: Reidel, 21–50.

—, 1986b. "The Triumphant Apocalypse and the Catastrophic Apocalypse," in *Nuclear Weapons and the Future of Humanity*, ed. A. Cohen and S. Lee. Totow N.J., 131–50.

—, 1987. "The Age of Reason versus the Age of Revelation. Two Critics of Tom Paine: David Levi and Elias Boudinot," in *Deism, Masonry, and the Enlightenment. Essays Honoring Alfred Owen Aldridge*, ed. J. A. L. Lemay. Newark, Del., 158–70.

—, 1988. *Millenarianism and Messianism in English Literature and Thought 1650–1800*. Leiden.

Priestley J., 1787. *Letters to the Jews*. Birmingham.

—, 1789. *Letters to the Philosophers and Politicians of France*. London.

—, 1794. *The Present State of Europe Compared with Ancient Prophecies*. London.

Rossi, P., 1985. *The Dark Abyss of Time*. Chicago.

Sandeen, E. R., 1968. *The Origins of Fundamentalism*. Philadelphia.

Tuveson, E. L., 1972. *Millennium and Utopia*. Gloucester, Mass.

Webster, C., 1975. *The Great Instauration: Science, Medicine and Reform, 1626–1660*. London.

Westfall, R. S., 1980. *Never at Rest: A Biography of Isaac Newton*. Cambridge.

Westminster Confession = Anon., 1641. *The Confession of Faith.... Composed by the Reverend Assembly of Divines Sitting at Westminster.* London.

Whiston, W., 1708. *The Accomplishment of Scripture Prophecies.* Cambridge.

Whiston, W., 1749. *Memoirs of the Life and Writings of Mr. William Whiston.* London.

Whittlesley, W. L., 1981. "Elias Boudinot," in *Dictionary of American Biography*, 2:477–78.

Symposium

Niels Bohr Centennial

Man as Spectator and Actor in the Drama of Existence

SHMUEL SAMBURSKY

The gradual clarification of the basic concepts of quantum mechanics is most clearly reflected in eight essays by Niels Bohr, written over a period of 30 years, from 1927 until 1957 (Bohr 1934; Bohr 1958). Among these essays there is a paper, originally published in *Studia Orientalia* (Copenhagen 1953) and entitled "Physical Science and the Study of Religions." At the opening of this paper Bohr stresses his deep indebtedness to his colleague, the orientalist Johannes Pederson, in honor of whose seventieth birthday he was contributing his comments on the relations between the epistemology of quantum physics and the philosophy of Oriental religions. I have not read Pederson's works, but I happen to be fairly well acquainted with the relevant books and papers of my friend Professor Toshimiku Izutsu, the eminent philosopher and historian of Chinese religions; although I am not a scholar in this field, I venture to make a few remarks on its connection with Bohr's philosophy of quantum mechanics.

The key words relating to this connection are repeated several times by Bohr. For example, in his 1929 paper he says, "The new situation in physics has so forcibly reminded us of the old truth that we are both onlookers and actors in the great drama of existence." In his 1937 paper, he tells us that "for a parallel to the lesson of atomic physics ... we must turn ... to that kind of epistemological problems with which already thinkers like Buddha and Lao-Tse have been confronted, when trying to harmonize our position as spectators and actors in the great drama of existence." Finally, in his 1954 paper, he refers to "old Oriental philosophy and its admonition never to

E. Ullmann-Margalit (ed.) The Scientific Enterprise, 263–267.

forget as we search for harmony in human life that on the scene of existence we are ourselves actors as well as spectators."

Indeed, the problem of "spectator and actor" emerged only with the development of atomic physics, which was concerned with the not negligible interaction between the observed object and the instrument of observation. On the other hand, classical physics, whose conceptual structure was brought to so marvelous a completion by Einstein, rests on the assumption that it is possible to discriminate between the behavior of material objects and the question of their observation. The principle of relativity, which represents the supreme level of classical determinism, compensates for the subjective element in the description of physical evidence by introducing observers in different states of motion, so that the problem of interference with the object of observation does not arise. In quantum theory, however, we are concerned with the relationship of evidence furnished by different experimental arrangements, among which every observer has a free subjective choice. The action of the measuring instruments implies mutual exclusion of the various kinds of information required for a complete description of the physical data, and the need to allow for the conditions under which experience is obtained calls directly for the complementary mode of description.

Again and again Bohr stresses the fact that no sharp separation between object and subject can be maintained, since the perceiving subject — the observer or spectator — also belongs to our mental content, and thus phenomena are *not* independent of our observations. In this connection Bohr cites a striking example, taken from daily life: the sensation which we experience when we attempt to orient ourselves in a dark room by feeling with a stick. When the stick is held loosely, it appears to our sense of touch to be an object. However, when it is held firmly, we lose the sensation that it is a foreign body, and our impression of touch immediately becomes localized at the point where the stick is touching the body under investigation.

Bohr's instructive illustration of the blurring of the boundary between subject and object implies two well-known particulars of the so-called "Copenhagen interpretation" of quantum theory, namely, that there exists a division — a cut — between the observer, including his measuring device, and the external object which he wants to observe; furthermore, this cut is to some extent arbitrary, and thus the addition of some part of the measuring device or the whole device to the object being measured cannot alter the final result of the experiment and its interpretation. One should not forget that the measuring device, by its very definition, must be in contact with the object observed; in other words, there is an interaction between the observer and his object. This interaction leads to uncertainty with respect to the behavior of the external world and to the establishment of a primary probabilistic explanation, in contrast to the deterministic one of classical physics.

These observations are certainly familiar to the reader. However, in this short summary I wish to throw into relief the close affinity of Bohr's ideas with the basic tenets of Taoism and Zen Buddhism. Lao-Tse and Chuen-Tse, ca. 300 B.C., developed the Tao philosophy, which opposed the Confucianism that preceded them by some 200 years. They denied the essentialism of Confucius, his belief that Reality consists of things solid and permanently fixed, of essences. According to Taoism, human language has the essential tendency to create distinctions that look as if they were "essential" boundaries between things. But Reality as pictured by language is distorted precisely because of the boundaries that it creates, for language fosters and supports essentialism; it forces many articulations upon the fluid and amorphous reality, whereas Reality itself has no such articulations. Taoism asserts that the fluid state of things on the level of daily experience is but a reflection of the metaphysical fluidity of Reality. Certainly, there *are* boundaries, but these boundaries themselves are blurred, they are characterized by fluidity, they are constantly changing. And since language fixes and stabilizes these ever-flowing boundaries, they become fossilized shapes, devoid of life. The real Reality, however, lies hidden beyond the "essences," and on a deeper level all things interpenetrate each other.

More than 1100 years after Lao-Tse, from ca. A.D. 850, the basic ideas of Taoism were elaborated by Zen Buddhism, whose philosophy emphasized the functional relationship between subject and object, between the knower and the known. Zen Buddhism centers round the concept of "relatio," in contradistinction to that of "substantia" in Confucian or also in Platonist-Aristotelian philosophy.

There is a very close correlation between the state of consciousness of the subject and the state of the objective world which the subject perceives. This correlation between subject and object is of an extremely subtle and dynamic nature, so much so that the slightest move on the part of the subject necessarily induces a change on the part of the object. Thus, the image of Man peculiar to Zen Buddhism transcends the bifurcation of subject and object. This metaphysical unity of all things, of the object and the observing subject, becomes manifest after a total transformation of the empirical ego; this was, for example, the basis of Cartesian dualism, where the spectator looked at things from the outside as a detached onlooker confronting the world of external objects.

However, the Zen Buddhist, instead of remaining an objective spectator looking at the world as something independent of him, finds himself in the very midst of the world, directly affecting the objects and being directly affected by them. Man himself *is* on the stage; he exists in the world, actively participating in the play. Existence, Being, is a *dynamic* field, the unbroken unity of the whole. Being is of a relational nature. It actualizes itself either as the subject — "I," "this" — or as the object — "that." "I" is not my empirical ego, but is a concrete actualization of the entire field of Being, it is ready at any moment to be assimilated from "this" into "that." In short, there is a dynamic relation between subject and object. The *subjective* dimension of Man is of decisive importance, as it determines the *objective* dimension in which Reality discloses itself to him; Reality becomes visible to us only at this extreme limit of our subjectivity. Since subject is subject only relative to the object and vice versa, externalization of the internal and internalization of the external lead in the final stage to the same result.

Returning again from Zen Buddhism to atomic physics, I want to quote some of Heisenberg's comments on the Copenhagen interpretation of quantum theory. He emphasizes the *subjective* element in the description of atomic events, since the measuring device has been constructed by the observer, and we have to remember that what we observe is not nature itself but nature exposed to our method of questioning. In our scientific relation to nature, our own activity becomes especially important when we have to deal with parts of nature into which we can penetrate only by using the most elaborate tools. Elsewhere, Heisenberg puts it this way: "Natural science is not Nature itself, but a part of the relation between Man and Nature, and therefore is dependent on Man."

In conclusion, I should like to add a few words on a relevant aspect of the methodology of physical science. Pauli, in his remarks entitled "Relativity and Science" at the conference dedicated to the fiftieth anniversary of relativity (Bern 1955), said that "Einstein has shown us again and again, how the physicist, without the help of fixed rules, must learn to swim in a boundless sea of ideas, to which he might be inspired by an equally boundless sea of empirical material, which, however, cannot be logically derived from it." Indeed, Einstein himself, in a talk about the methodology of theoretical physics in 1930, polemicized against Newton's belief that the basic concepts and laws of his system — mass, inertia, force — can be derived from experience alone. Only general relativity, Einstein says, led to a clear recognition that Newton's view is not correct, and that every attempt at a logical derivation of the basic laws of mechanics from elementary experiences must fail.

It is obvious that the development of the quantum theory is a still more convincing illustration of Einstein's words than that of the relativity theory.

The formidable structure of quantum mechanics was not erected by decades of pure experiments *alone*; the whole content of the system crystallized after a long and laborious process of *interpretation* of those experiments, which finally formed the mental fabric of quantum physics. A few years later, in his essay on "Physics and Reality" (1936), Einstein again emphasized that it is wrong to gloss over the logical independence of scientific concepts vis-à-vis our sense perceptions. He used a witty picture from everyday life in order to illustrate the complex relation of theory and experiment: The question, he said, is not that of an *extraction*, like the relation of meat soup to beef, but rather that of a *correlation*, such as the relation of the cloakroom number to the overcoat.

References

Bohr, N., 1934. *Atomic Theory and the Description of Nature*. Cambridge: Cambridge University Press.

Bohr, N., 1958. *Atomic Physics and Human Knowledge*. New York: Wiley.

Einstein, A., 1936. "Physics and Reality," *Franklin Institute Journal* 221:349–82.

Pauli, 1955. *Niels Bohr and the Development of Physics*. New York: McGraw-Hill.

Bohr's Response to Einstein, Podolsky, and Rosen

DAVID Z. ALBERT

In celebration of Niels Bohr's centennial, I would like to argue with him. It would be presumptuous of me, I think, to praise him, since he is hardly in any need of praise from people like myself.

Let us recall and reconsider Bohr's reaction to the famous argument of Einstein, Podolsky, and Rosen of 1935. Their paper began with the statement of a proposed sufficient condition that a certain physical property be considered an "element of reality," which ran like this:

> If, without in any way disturbing a system, we can predict with certainty the value of a physical quantity, then there exists an element of physical reality corresponding to this physical quantity.

That sounds commonsensical enough. It is surely consistent with what the man in the street ordinarily thinks about physical reality, and (as the authors point out) it is true of the picture of physical reality to which we have become accustomed in classical physics.

Einstein, Podolsky, and Rosen (henceforth referred to as E.P.R.) suppose that "physical reality," whatever precisely that may ultimately come to mean and whatever form it is ultimately found to take, must certainly be what physicists have come to call "local." That physical reality must be local means that what occurs here can have no immediate effects elsewhere; more precisely, it means that what happens at one space-time point *A* cannot possibly affect the conditions at another such point *B*, unless *A* and *B* can be connected by

E. Ullmann-Margalit (ed.) The Scientific Enterprise, 269–272.

some causal physical process, a process which can propagate through space no more rapidly than a light beam.

These are the assumptions of E.P.R. Their argument runs as follows. They consider circumstances in which two particles (1 and 2, say) are known to be correlated in a number of their physical properties, so that by measuring property P of particle 1 one can ascertain the value of property P for particle 2, and similarly for some other property Q. It turns out that such simultaneous correlations can be arranged even in cases where P and Q are quantum-mechanically incompatible observables. Consider then (as E.P.R. do) such a case. According to the conventional understanding of quantum mechanics, P and Q of particle 2, say, cannot both be taken simultaneously to have a physical meaning; they cannot both be considered simultaneously to have real and well-defined values; they cannot both be intelligibly described. But note (say E.P.R.) that, once these two particles are correlated, they can be separated any distance in space, and the separation in no way impairs our ability to infer with certainty the properties of particle 2 from those of particle 1. Furthermore, the locality assumption implies that measurements of P or Q for particle 1 can in no way affect any of the properties, can in no way affect the "reality," of particle 2. Now, although P and Q cannot *both* be measured for particle 1 (they are, we recall, incompatible), *either P or Q could* be measured for particle 1. Since by measuring P for particle 1 we can predict with certainty the value of P for particle 2 without (given the locality assumption) in any way disturbing particle 2, it follows that P of particle 2 must be an element of reality. The same argument applied to Q will establish (although P and Q cannot *both* be measured for particle 1, but merely because either *one* of them can) that Q is *also* an element of the reality of particle 2.

Consequently, the quantum-mechanical description of the world must be incomplete (since it does not contain formal elements corresponding to every element of the reality of particle 2, in our example); and E.P.R. envisage that waiting somewhere to be discovered, there must exist a complete, local, deterministic, realistic, objective picture of nature.

The most striking thing about Bohr's reaction to E.P.R. is that it has such an overtly philosophical tone. Bohr does not take issue with E.P.R. on the level of physical considerations. Rather, he makes it clear from the outset that in his opinion their argument "discloses an essential inadequacy of the customary viewpoint of natural philosophy for a rational account of the physical phenomena of the type with which we are concerned in quantum mechanics." More particularly, Bohr claims that what is necessary for the resolution of the problems raised by E.P.R. is nothing less than "a radical revision of our attitude towards physical reality."

Precisely what kind of revision Bohr had in mind is an obscure and difficult matter which, I must confess, I have never entirely understood; but let us concentrate here on one particularly clear and simple component of that revision. The E.P.R. argument involves counterfactuals. Although *P* and *Q* cannot *both* be measured for particle 1, the argument hinges on statements about what would happen in the event that *P* were measured and what would happen in the event that *Q* were measured. Bohr, in his response, proclaims (without, it should be noted, any semblance of an argument) that such statements are henceforth simply to be disallowed. Either *P* or *Q*, but not both, can be measured for a given system. If *P* happens to be measured, then questions about what would have happened in the event that *Q* had been measured are to be regarded as "unaskable" and unintelligible; and, of course, vice versa. He speaks of "*renunciation* in each experimental arrangement of one or the other of the aspects of the description of physical phenomena," and of "discrimination between different experimental procedures which allow of the unambiguous *use* of complementary classical concepts."

Consider, for a moment, what Bohr is proposing here. Counterfactual statements (statements like, say: "*If* I had dropped this pen five minutes ago, which in fact I did not, it *would* have fallen to the ground") are by their very nature experimentally unverifiable; and, as Bohr points out, statements about measurements of both *P* and *Q* for a single particle must necessarily be counterfactual statements, and so must necessarily lie beyond the possibility of experimental verification. But that in no way means (think of the pen!) that such statements are *unintelligible*!

The assumption that such statements are for some reason unintelligible is a dogma, which fell upon physics like a divine pronouncement and harmed it, in my opinion. Such questions as Bohr had disallowed were, for a long time, simply not raised in physics; and the striking advances in our understanding of the quantal world which the resolute asking of those questions finally produced, decades later, was denied us, for no good reason, for many years.

It was John Bell who, finally, resolutely asked those questions. Ironically, the famous theorem that resulted supports Bohr's point of view far more than it does that of E.P.R. It turns out (this is the content of Bell's result) that Einstein's dream of an objective, deterministic, local theory consistent with the experimental facts cannot possibly come true. More particularly, Bell was able to show that, no matter what the correct theory of the world is, the world does *not* have the property of locality; and since E.P.R. relied on the locality assumption their argument simply fails.

In the final analysis, Bohr's worst enemy here was himself. His intuition was, if such a thing is possible, too strong. He saw that the E.P.R. argument was flawed, but could not say precisely in what way. He responded to the argument with dogma; and the dogma was not only unsupported, but delayed the discovery of a rigorous refutation of the E.P.R. argument.

It seems to me that one of Bohr's most profound and original powers, something that lay very near the center of his shining genius, served him badly in this instance. He took physics seriously. He thought that great things, the deepest and most important truths, were at stake there, and he had a radical, adventurous, unblinking, and unfearing zeal to read of those truths in the book of physical nature. He was on the lookout for immense things; he was determined to find them. Sometimes he was ready to jump to conclusions. Think, for example, of complementarity, which we have been considering here. Bohr supposed that complementarity would prove to be so basic and so general a characteristic of our thinking about the world that it would be taught, along with basic arithmetic, say, in the elementary schools of the future. Bohr saw complementarity in biology, he saw it in the mind-body problem; he saw it everywhere. I have heard that he once said: "A trivial truth is one whose negation is false; a profound truth is one whose negation is also true!"

Looking back, this all seems somewhat premature. The E.P.R. argument failed, in the end, not because it raised improper questions, as Bohr would have it, but because it proceeded from a false assumption: the locality assumption. As a matter of general policy, one should be extremely reluctant to ban the posing of questions. Indeed, a number of those, besides Bell, who have been honest and courageous enough to pose the questions which Bohr disallowed (I am thinking here particularly of David Bohm and the late Hugh Everett) have produced very intriguing and astonishing suggestions as to how those questions might finally even be *answered.* Bell, who is determined not to repeat Bohr's mistake, has taken much interest, and generated much interest, in those suggestions. I think he is doing the right thing.

The Genesis of Bohr's Complementarity Principle and the Bohr-Heisenberg Dialogue*

MARA BELLER

Introduction

The aim of this paper is to reconstruct the critical stages in Bohr's thought, which led him in 1927 to formulate his complementarity principle as a response to Heisenberg's uncertainty paper (Heisenberg 1927). I will argue that Heisenberg's uncertainty principle, and the ensuing clash between Bohr and Heisenberg, played a central role in the development and elaboration of Bohr's interpretation of quantum physics. My paper "complements" existing accounts (Jammer 1966; Jammer 1974; MacKinnon 1982) which relate Bohr's notion of complementarity to the problem-complex of the infamous Bohr-Kramers-Slater paper (Bohr, Kramers, and Slater 1924), but seem to underestimate the magnitude and the nature of Bohr's dialectical reaction to Heisenberg's ideas.

To be sure, Bohr's initial grappling with wave-particle duality, as well as his realization that a causal space-time description in the atomic domain might not be possible, can be traced to the ideas expressed in the Bohr-Kramers-Slater paper. Yet, Bohr's hesitant, almost resigned statements after he had learned the results of the Bothe-Geiger experiments were far removed from the confident, full-fledged and closely-knit elaboration of complementarity in his Como lecture (Bohr 1928a).

* This paper was written before J. Heilbron's (1988) penetrating analysis of the diffusion of the Copenhagen spirit became available to me.

E. Ullmann-Margalit (ed.) The Scientific Enterprise, 273–293.

The "critical dialogue"[1] with Heisenberg played an indispensable role in the birth of Bohr's complementarity principle, not in its initial conception, but rather in its maturing stage. In particular, the central pillars of the complementarity principle — wave-particle duality, the definition of concepts versus observational possibilities and the indispensability of classical concepts — all crystallized in response to Heisenberg's position.

Heisenberg was not the only physicist Bohr argued with — the Como lecture was at the same time a dialogue with Schrödinger, Einstein, Dirac and Born.[2] In fact, almost every sentence in Bohr's Como lecture is an implicit argument with one of these key figures. Viewed from this perspective, Bohr's presentation of complementarity (Bohr 1928a) becomes at once more transparent and more profound.

This paper also attempts to throw some light on a puzzling question: the source of the inconsistent mixture of classical and quantum concepts within the orthodox interpretation of quantum physics. Recently, I argued (Beller 1986) that, in order to be able to trace the source of such inconsistencies, we must supplement the description of conceptual evolution with an analysis of "external" socio-psychological factors. The brief discussion of the ideological role of complementarity at the end of this paper suggests what type of analysis I have in mind.

Heisenberg's Uncertainty Paper

The problem of a physical interpretation of the new quantum theory became pressing after Schrödinger had proved the equivalence of the matrix-mechanical and wave-theoretical formalisms (Schrödinger 1926). Both theories had previously been used as calculational devices, but their physical significance was unclear. The necessity of an adequate interpretation along Göttingen-Copenhagen lines became especially acute after Schrödinger's attempts to provide a semiclassical interpretation had met with considerable enthusiasm in wide scientific circles (Heisenberg 1971, 73). Schrödinger's antipathy to "damned quantum jumping" was fully revealed in his adamant opposition to Bohr and Heisenberg during his stay in Copenhagen, in September 1926 (ibid.). An extremely intense search for a way to counter Schrödinger's "heretical" ideas then occupied both Heisenberg and Bohr throughout the winter of 1926. The first fruit of this search was Heisenberg's uncertainty paper (Heisenberg 1927), written in February 1927. I shall now summarize the central points of Heisenberg's argument,[3] treating Bohr's reaction to each of these points in the following sections:

1. The central aim of Heisenberg's uncertainty paper was to provide an

interpretation of the new quantum mechanics that would render that abstract theory intuitable, or visualizable (*anschaulich*). The earlier radical philosophy of matrix mechanics had renounced *Anschaulichkeit* and had eliminated the space-time description within the atom. A pictorial, kinematic corpuscular description was considered not only unfruitful, but downright sinful (unobservable entities should be eliminated from physics). The atom thus became a "black box," which was in principle not capable of being cracked. After the appearance of Schrödinger's wave mechanics, however, with its familiar classical concepts and wide intuitive appeal, Heisenberg redefined *Anschaulichkeit* and restored the usual space-time container to the atom's interior.

The earlier version of matrix mechanics forbade a particle description of the atom's interior; the only legitimate physical concepts were electromagnetic ones, connected with the "virtual oscillator" model. Initially, not only was the orbital motion within the atom eliminated, but the concept of the electron's position was also rendered meaningless. In his uncertainty paper, however, Heisenberg resurrected visualization of the electronic positions within an atom, describing the atom as a system of individual particles that obey statistical laws.

Moreover, the particulate description was an exclusive one — Heisenberg, who perceived Schrödinger's wave-mechanics as a competing system, was avoiding wave-theoretical concepts at any price. In particular, he provided an incorrect analysis of the gamma-ray thought experiment, relying exclusively on corpuscular aspects of matter and radiation. Heisenberg later recalled that he had "wanted to forget about wave packets and about waves," hoping that a "mathematical scheme would be clever enough to deduce the waves from the particulate approach, without physicists' thinking about it."[4] Heisenberg also put forward forceful and convincing arguments against Schrödinger's attempts to represent particles as wave packets that keep their identity.

2. Heisenberg's avoidance of wave-theoretical concepts in his uncertainty paper was connected with his belief that Schrödinger's competing system was merely a convenient mathematical tool for the easy calculation of matrix elements. Heisenberg considered Schrödinger's mathematical formalism as subservient to the matrix theory and as lacking independent physical significance. Moreover, unlike Schrödinger's wave-theory, matrix theory, interpreted in kinematic corpuscular terms, is well suited to represent the undeniably discontinuous aspect of quantum phenomena. Schrödinger's interpretational attempts are misleading and deviate from the path initiated by the founders of quantum theory — Bohr and Einstein.

3. Despite Heisenberg's emphasis on particles, he was surely no "naive

realist. In fact, he redefined "reality" and basic physical concepts operationally, in order to make sense of the highly abstract and unintuitive formalism of the Dirac and Jordan transformation theory. For Heisenberg, in the true Göttingen spirit, "mathematics knows better than our intuition"[5] and ultimately defines the meaning of observable phenomena. In the uncertainty paper Heisenberg took an extreme operationalistic position (the meaning of theoretical terms is equivalent to the operation of their measurement). By operationally redefining the meaning of familiar classical concepts, Heisenberg succeeded in providing an intuitive physical interpretation of the new abstract formalism. All classical concepts, he concluded, can be retained in the new theory, if one endows them with strict operational meanings and excludes the simultaneous rigorous use of conjugate variables.

"Black Box" and Wave-Particle Duality

Bohr's realization that a causal space-time description might not be possible in the quantum domain has its roots in the Bohr-Kramers-Slater theory (Bohr, Kramers, and Slater 1924). In the original paper, the wave description of radiation demanded the renunciation of regular kinematic description within the atom (the virtual oscillator model), in addition to the renunciation of conservation laws (causality) for the individual processes of emission and absorption of radiation. When the Bothe-Geiger experiments (Bothe and Geiger 1925) proved that conservation laws do indeed apply to individual processes and require a particulate description of radiation, a dilemma arose: how does one reconcile the wave space-time description with causality (conservation laws)?

Bohr realized, apparently, that an exclusively particulate non-statistical picture of light is impossible — "photons" do not travel along regular space-time trajectories. On the other hand, a continuous wave description can be related to discontinuous energy changes only through probability arguments. This problematic situation thus prompted Bohr's initial understanding that, although space-time and causality, or waves and particles, are both necessary, they cannot be applied simultaneously to a given physical situation. After the Bothe-Geiger experiments (ibid.), Bohr subjected both the applicability of the familiar classical concepts and the classical ideal of explanation in terms of spatiotemporal pictures to a most serious scrutiny. And after Heisenberg's painfully abstract matrix mechanics had appeared, Bohr confided to his philosophy teacher Höffding that he had become "more and more convinced of the need of a symbolization if one wants to express the latest results of physics" (cited in Jammer 1966, 347).

Yet Bohr's aknowledgment of the need for symbolizing was, I believe, reluctant and ambivalent. Indeed, just a few months after the discovery of the new matrix mechanics, Bohr expressed his optimism concerning the applicability of mechanical pictures in relation to the Goudsmit and Uhlenbeck idea of the spin of an electron:

> This idea allows a remarkable way to overcome the apparently grave difficulties in the spectral theory which have puzzled us so much in later years. In fact, these difficulties gradually created that scepticism as to the unambiguous use of mechanical pictures even for qualitative conclusions.... It looks now that at any rate in the qualitative way characteristic for the correspondence principle we shall be able by means of such pictures supplemented with the new idea to account comprehensively for the building up of atoms.[6]

In a public lecture[7] given at that time, Bohr hardly mentioned the new matrix mechanics. He must have been aware of the tremendous conceptual difficulties in the new theory.[8] Nor could he have been happy that Heisenberg, despite having known that it was the correspondence principle that had enabled him to construct the new formalism, had not mentioned Bohr's correspondence principle at all in his original paper (Heisenberg 1925), presenting his new advance as flowing from the philosophy of elimination of unobservables.[9]

However, that philosophy was a short lived one and did not lead to a consistent interpretation of the new quantum theory (Beller 1985b). In his uncertainty paper Heisenberg implicitly abandoned his earlier philosophy and restored the exiled unobservables. From that time on, Heisenberg described and legitimized his procedures in his reinterpretation paper (Heisenberg 1925) as having derived from the most comprehensive and rigorous use of the correspondence principle, in contrast to the justification of these procedures given in the paper itself!

Heisenberg's failure to provide a self-sufficient interpretation of the new formalism based on new principles, dissociating the "context of discovery" and the "context of justification," must have impressed Bohr deeply. The correspondence principle and the familiar classical concepts had played a fundamental role in the construction of the new quantum theory — hence they must also play an indispensable role in interpreting it. Heisenberg's failure to provide a satisfactory alternative could not but strengthen Bohr's position.

In addition, the need to counter Schrödinger's threat forced both Bohr and Heisenberg feverishly to explore the problem of interpreting the new matrix-mechanical formalism in terms of familiar space-time concepts (Beller 1985b). This was indeed the crux of their quest for an interpretation in 1926. O. Klein's recollections[10] confirm this point: "I think they [Heisenberg and Bohr] probably had discussed the problems of how to introduce space and time and all these things into these very abstract formulas." The result of the search was Heisenberg's cracking of the "black box" and his description of the interior of the atom in terms of space-time concepts of individual particles (Heisenberg 1927).

Yet Bohr, despite his reluctance to abandon visualization, was not as hasty as Heisenberg to restore the space-time container inside the atom. Physicists' knowledge of the interior of the atom is, and always will remain, indirect (Bohr 1928a). Bohr was acutely aware that in quantum theory, unlike classical theory, there is a profound difference between free and bound particles. While free electrons can be treated as mass points, bound ones defy this simple visualization. Schrödinger's eigenfunctions had provided an adequate and straightforward representation of the stationary states of an atom, but the "consistent application of the idea of stationary state excludes ... any specification of regarding the behavior of the separate particles in the atom" (ibid., 581). Stationary states are characterized by definite momentum and energy values; their space-time coordinates must therefore be undetermined. In fact, "on the whole it would not seem justifiable, in case of interaction problem, to demand a visualization by means of ordinary space-time pictures" (ibid., 580). Since the space-time description is not applicable to an atom in a definite stationary state, "an immediate utilization of the concept of energy in connection with the classical idea of the potential energy of the system is excluded" (ibid.). The idea of stationary state is therefore irreconcilable with visualizing intra-atomic particles separately in a regular space-time container.

What must have further impressed Bohr at this time was Heisenberg's failure to provide a corpuscular interpretation for the interior of the atom, despite his repeated attempts to do so after Bohr had pointed out to Heisenberg his mistake in the analysis of the gamma-ray experiment. In his analysis Heisenberg treated both photons and electrons as regular point particles, arguing that a photon colliding with an electron transfers to the latter a discrete and uncontrollable amount of energy (Compton recoil) (Heisenberg 1927).

The more precisely the position of the electron is determined, the greater the uncertainty in the discontinuous change of the electron's momentum. Yet Compton recoil would not lead to indeterminacy, but to perfectly calculable changes: the momentum change of the electron could be calculated exactly from the conservation laws for energy and momenta and from a knowledge of the energy and momentum of the colliding photon.

The cause of Bohr's severe criticism of — and opposition to — the uncertainty paper was not a priori metaphysical disagreements (as Heisenberg's published recollections suggest), but rather Heisenberg's mistakes. Bohr did in fact point out Heisenberg's mistakes publicly in his Como lecture — Compton recoil "could not prevent us from ascribing accurate values to the space-time coordinates, as well as to the momentum-energy components before and after the process" (Bohr 1928a, 572–73). Klein's recollections confirm that the Heisenberg-Bohr clash over the uncertainty paper followed not so much from philosophical disagreement, "it was more the technical point that Heisenberg did not discuss these experiments correctly" (ibid.). Besides, the analysis of Compton recoil applies rigorously to free and not to the bound electrons — Bohr must have argued this point aggressively enough to force Heisenberg to concede it explicitly, in a postscript to his uncertainty paper (Heisenberg 1927).

If Heisenberg's analysis of the gamma-ray experiment was not really applicable to intra-atomic electrons, he had clearly failed to provide an adequate interpretation of the quantum domain in particulate terms. No wonder Bohr considered the publication of the paper premature and urged Heisenberg not to submit it for publication. Heisenberg and Bohr argued this point again and again, and the debate always centered around the question of whether the wave or the particulate aspects of radiation and matter were essential in the analysis of Heisenberg's thought experiment: "Bohr emphasizes that the diffraction of waves is essential in the gamma-ray microscope, I emphasize that the light quantum theory and even the Bothe-Geiger experiment are essential" (Heisenberg's letter to Pauli on April 4, 1927, in *Pauli's Scientific Correspondence*). They argued this point until Heisenberg, desperate at his inability to counter Bohr's correct arguments, burst into tears: "I did not know exactly what to say to Bohr's argument to begin with and so the discussion ended with the general impression that now Bohr again has shown that my interpretation is not correct."[11]

Thus Heisenberg's repeated failure to provide a satisfying particulate kinematic interpretation without recourse to waves must have considerably strengthened Bohr's opinion that wave concepts are unavoidable. He therefore concluded that Schrödinger's wave-superposition notion for the interior of the atom seems to be much closer to the truth. But nor is the wave concept alone sufficient: In his Como lecture Bohr repeated Heisenberg's objections to Schrödinger, citing Heisenberg's uncertainty paper. After all,

Bohr could hardly sympathize with Schrödinger's aspirations to do away with particles and "damned quantum jumping" — concepts on which his entire approach to the atomic domain was based.

Bohr thus became convinced that a one-sided description (either in terms of particles or in terms of waves) is not possible. The atom is a "black box" and all knowledge of its structure is indirect. This knowledge actually derives from two sets of experiments — experiments on the atom's radiation and experiments on atomic collisions. While collisions are most naturally described in terms of particles, Bohr continued to believe that radiation is basically of a wave nature.

The concept of a stationary state, described in wave terms, is inconsistent with the particulate kinematic description. Yet it has as little reality as the concept of an individual particle; both concepts are abstractions, whose applications depend on the nature of the problem and the possibility of consistently using such mutually exclusive concepts.

It is worth pointing out that Max Born's original probabilistic interpretation of the wave function (Born 1926a, 1926b) was also in accord with Bohr's emphasis on the distinction between free and bound particles. Treating the atom as a "black box" in his original papers, Bohr never talked about the probability of the position of an intra-atomic electron. If the wave function determines the probability that a free electron will move in a certain direction in space, the wave function for the atom's interior determines the probability that it will be in a certain stationary state, or the probability that it will undergo a certain transition from one stationary state to another.

Unlike Born and Heisenberg, however, Bohr did not consider these waves to be "immaterial," "ghost waves" whose essence is of a mathematical, probabilistic nature. He agreed with Schrödinger that there must be something very real about these waves inside an atom. Heisenberg himself, when the fervor of his passionate controversy with Schrödinger had cooled down, actually conceded that Bohr had been right on this point all along:

> Actually, we need not speak of particles at all. For many experiments it is more convenient to speak of matter waves, for instance, of stationary matter waves around the atomic nucleus ... there the wave picture comes much nearer to the truth than the particle picture. Therefore, Bohr advocated the use of both pictures which he called "complementary to each other" (Heisenberg 1959, 49).

Physical Significance of Schrödinger's Theory

At the time Heisenberg was conceiving his uncertainty principle and Bohr was pondering complementarity, the former was not willing to grant Schrödinger's formalism any physical significance. This position was in fact an implicit assumption behind Heisenberg's uncertainty paper. It was a common stand among Göttingen physicists; they related to Schrödinger's theory merely as a mathematical appendix of matrix mechanics (see Beller 1985a, 1985b). Even when they used Schrödinger's formalism, they stated that it was only a convenient calculating technique.

Bohr never approved of this state of affairs. According to Klein, Bohr actually scolded F. Hund for expressing such sentiments in his paper on molecular spectra:

> This was one of the early Göttingen papers that really used the Schrödinger equation, and it was getting to be customary to say "because it's easier, or better fits this problem, I shall use wave functions, but of course only for the convenience of the mathematics." ... But he [Hund] says that Bohr said to him ... its general bearing was: that really isn't right ... they [wave functions] have some physical meaning, they are physical in some sense that this sort of statement does not admit.[12]

Unlike Heisenberg, who wanted to eliminate Schrödinger's physical ideas, Bohr wanted to incorporate Schrödinger's physical insights into his own framework.

Why was Bohr predisposed towards Schrödinger's ideas?[13] One of the main reasons, I believe, was what he considered to be Schrödinger's resurrection of Bohr's own concept of a stationary state. This concept was Bohr's most original, most brilliant and most fruitful contribution to physics. It not only explained the stability of atoms, which is incomprehensible in the classical framework, but also demystified the complex spectral "music."

Yet matrix mechanics suppressed the concept of the stationary state: not only was there no mathematical description of an atomic state by the matrix-mechanical method, but the concept of a sequence of stationary states with increasing energies also lacked physical significance (Beller 1983). This inability of the matrix approach to describe an atom in a given stationary state was considered by some physicists to be a serious drawback (ibid.).

In Bohr's opinion, Schrödinger's wave mechanics provided just what was needed: an adequate representation of stationary states. Schrödinger's approach "has proved to be of decisive importance for the great progress in atomic physics during the last years. Indeed, the proper vibrations of the Schrödinger wave equation have been found to furnish a representation of the stationary state of an atom meeting all requirements" (Bohr 1928a). Moreover, the continuous distribution of charge and current, which

Schrödinger associated with the solution to the wave equation, adequately represented the electrical and magnetic properties of the atom in the corresponding stationary state.

Clearly, Schrödinger's wave approach, far from being a mere mathematical device, contained profound physical truth. That Bohr indeed saw this aspect of Schrödinger's theory as its central asset is revealed in his letter to Kronig of 28 October 1926:[14]

> [A] few weeks ago we had a visit of Schrödinger, which gave rise to much discussion regarding the physical reality of the postulates of the atomic theory . . . Indeed they [Schrödinger's methods] offer a most welcome supplement to the matrix mechanics in allowing to characterize the stationary states separately.

Bohr's opinion that Schrödinger's wave functions described adequately separate stationary states inevitably implied that no corpuscular-kinematic description of the interior of the atom in terms of individual particles is possible: "In problems where a description of this behavior is essential, we are bound to use the general solution of the wave equation which is obtained by superposition of characteristic solutions" (Bohr 1927a, 581). Hence Heisenberg's hopes "to avoid waves" were fundamentally doomed. Moreover, it is Schrödinger's wave approach that explains why a description of the "orbit" of the electron in the atom is impossible: "From characteristic vibrations with only a few nodes no wave packages can be built up which would even approximately represent the motion of a particle" (Bohr 1928b, 113).[15]

Definition Versus Observation

Bohr's conception of stationary states was problematic from the very start, and the attendant conceptual difficulties occupied him long before his complementary principle. Not only did the instantaneous transitions between the different stationary states imply that electrons were "clairvoyant"; they were also inconsistent with the observed width of the spectral lines. The problem of the dependence of the breadth of spectral lines on the lifetime of stationary states was discussed by Bohr and Slater in connection with the Bohr-Kramers-Slater theory (Slater 1975). It was common physical knowledge that a wave train of finite length has a frequency spectrum that is not strictly monochromatic, but has a frequency breadth of the order of magnitude of $1/T$, where T is the time during which the wave train is emitted. Experiments in which the breadth of spectral lines was measured indicated that radiation was emitted over a considerable length of time (not instantaneously), and this led to the assumption of continuous (albeit virtual) radiation during the lifetime of stationary states in the Bohr-Kramers-Slater theory. This realization also led to "the first instance of the indeterminacy principle — that a stationary state cannot be better defined than *h* divided by the life-time".[16]

This reasoning, based on the knowledge that wave groups (identified with particles) are "necessarily accompanied by a lack of sharpness, in the definition of period and wave-length," was central to Bohr's analysis of the uncertainty principle in his Como lecture (Bohr 1927). The uncertainty stems not from uncontrollable changes during measurement (Heisenberg), but from an analysis of the definition of familiar physical concepts (Bohr). This disagreement about the underlying physical basis of the uncertainty principle was another major confrontation between Heisenberg and Bohr. But here, again, what was at issue was not simply a matter of different philosophical predispositions; Bohr had very good reasons not to be convinced by Heisenberg's stand.

Bohr, most likely, was aware that Heisenberg's initial conception of uncertainty for individual particles was based on an analysis of the definition of wave-theoretical concepts in Bohr's spirit. As Heisenberg wrote to Pauli on 28 October 1926, there is no sense in speaking of a monochromatic wave at a definite instant or in an extremely short time interval; but if one supposes that the line is not too sharp and the time interval not too short, the idea begins to make sense. Analogically, one cannot talk about the position of a particle moving at a definite velocity. Yet when one does not apply these concepts too rigorously, it may well make sense (*Pauli's Correspondence*, 350).

However, quite apart from the actual genesis of Heisenberg's uncertainty principle, Bohr clearly realized that Heisenberg had failed to give an adequate interpretation of the uncertainty principle in terms of possibilities of observation. As Heisenberg's analysis of the gamma-ray thought experiment was erroneous, he simply failed to deduce uncertainty from the changes occurring during the measurement process. The correct analysis of the experiment is based on diffraction arguments about the resolving power of the microscope — the possibilities of observation depend upon the definition of the wave-theoretical concepts, and not upon uncontrollable changes during measurement! This realization must have increased Bohr's confidence still further concerning the indispensability of wave-theoretical concepts.

Emphasis on the definition of concepts rather than on the possibilities of observation is really more appropriate for the "black-box" nature of the atomic interior. In his Como lecture Bohr pointed out several difficulties connected with the concept of observation in the atomic domain. According to Bohr, the emphasis in matrix mechanics on observability of the physical magnitudes was misleading to begin with — that matrix mechanics was called a calculus dealing with directly observable quantities hardly made sense for Bohr, for whom observing necessarily meant observing in the usual space-time: "The matrix mechanics has often been called a calculus with directly observable quantities. It must be remembered, however, that the procedure described is limited just to those problems in which in applying the quantum postulate the space-time description may largely be disregarded, and the question of observation in the proper sense therefore placed in the background" (Bohr 1928a, 577).

Though Bohr demonstrated at several points in his Como paper that the possibilities of observation are compatible with the possibilities of definition of physical concepts, he was by no means willing to base the definition of physical concepts on the procedure of their observation, as Heisenberg had done in his uncertainty paper. Perhaps Bohr also realized that Heisenberg had not really succeeded in redefining the regular physical concepts operationally. Take, for example, Heisenberg's operational definition of a particle's velocity: "If one wants to measure the velocity of an electron in an atom at a particular instance, the nuclear charge and the forces from the other electrons must suddenly disappear, so that a force-free motion results" (Heisenberg 1927). In order to measure the velocity of an electron in force-free motion one has to "illuminate the object with red light, and discover its velocity through the Doppler effect on the reflected light" (ibid.). Aside from Heisenberg's not having hinted at the kind of experiment, however hypothetical, that could suddenly eliminate nuclear and electronic forces, the procedure of determining velocity through calculations using the Doppler effect could hardly be considered a strictly operational one.

Some physicists realized this. Kennard, for example, criticized Heisenberg's operationalism (Kennard 1928). Kennard spent some time in Copenhagen in 1927, and he most likely discussed this central interpretative issue with Bohr. And the drawbacks of Heisenberg's position must have convinced Bohr that the issue was not one of operationally redefining familiar physical concepts, but rather one of analyzing the limits of applicability of already (classically) defined concepts within a new domain of experience. This implicit argument with Heisenberg contributed to the formulation, articulation and consolidation of Bohr's position concerning the meaning of physical concepts.

Heisenberg's failure to provide a comprehensive and consistent operational foundation for the new interpretation of physics was at the same time a failure to redefine *Anschaulichkeit* in his own terms. For Bohr, with his direct, down-to-earth physical intuition, Heisenberg's sophisticated redefinition of *Anschaulichkeit* must have seemed an illegitimate move. This was, in fact, one of the central issues in the clash between Heisenberg and Bohr over the uncertainty paper. As Heisenberg informed Pauli (*Pauli's Correspondence*, 395) there were essential differences of taste, or opinion ("wesentliche Geschmackunterschiede"), over the word *anschaulich*. Heisenberg's failure to furnish a new and adequate operational definition of the basic physical terms only strengthened Bohr's conviction that these terms have a well-determined meaning prior to and independent of any particular experiment in the atomic domain. Thus, Bohr concluded, the basic meaning of physical terms is to be elucidated by their use in everyday language and in the language of classical physics, which is a natural extension of ordinary language.

For Bohr, as for many interpreters of quantum physics after him, observation through laboratory apparatus necessarily meant observation in terms of familiar space-time concepts. Actually, Bohr was explicit about this: "Strictly speaking, the idea of observation belongs to the causal space-time way of description" (Bohr 1928b, 102). The central pillar of quantum orthodoxy — the indispensability of classical concepts — is based on this point.

Yet when Bohr insisted that the original version of matrix mechanics was not really a theory of observables, because it did not operate with space-time concepts, he was not doing justice to the bold, original philosophy of that theory. For the observables of matrix mechanics were indeed not positions and momenta, but frequencies and intensities of emitted radiation. The colors of the spectrum, the marks on a photographic plate, do not necessitate a full-fledged Newtonian container. In fact, this was the most exciting philosophical message of matrix mechanics — agencies of observations do not have to be described in regular kinematic space-time concepts.

The necessity of classical concepts is not as self-evident as many philosophical accounts in the Copenhagen spirit would suggest. Dirac, who felt reservations about complementarity throughout his life,[17] expressed his doubts to Bohr from the very beginning:

> Who didn't quite agree with Bohr at first was Dirac ... he made some kind of observation philosophy which had to do with permanent marks. Bohr objected a lot to that because it ... is not the essential thing in the observation. But Dirac made that, so to say, as the essential thing ... because one part of Bohr's view was that you had to, and you can,

> use the whole classical theory in describing observation. And I think that was not near to Dirac's mind at that time.[18]

The Necessity of Classical Concepts and the Ideological Role of Complementarity

The doctrine of the indispensability of classical concepts forms the backbone of the Copenhagen interpretation of quantum physics. It necessitates a duality of description based on waves and particles, or causality and space-time pictures. For if classical concepts are not indispensable, they might be replaced by another conceptual scheme that avoids this "schizophrenic"[19] duality or splitting of one world into two irreconcilable halves — waves and particles, or the quantum and classical domains. It is this central pillar of quantum orthodoxy that is most responsible for what Capek has aptly called "an oscillation between the Newtonian subconscious and its conscious rejection" (Capek 1961), or what Fox called the failure to construct a novel, non-classical paradigm "matching the radicality of the acquired knowledge" (Fox 1979) of quantum physics.

Bohr's insistence on the indispensability of classical concepts merits a separate analysis. A few remarks nonetheless follow.

According to Bohr, the inevitability of classical concepts follows from the fact that they constitute a necessary extension of intuition. Bohr argued this point in a letter to Schrödinger, as early as May of 1928.[20] This proximity of classical concepts to direct intuition is not, in my opinion, easily defensible, and it may well be wrong. Is not common sense Aristotelian, rather than Newtonian? Is the concept of universal, absolute time, which applies simultaneously to the entire world, intuitive at all? Is time-experience not local? And if so, is the Einsteinian concept of multiplicity of time rates for different observers any less intuitive than the Newtonian conception of time? Is common sense perhaps simply reeducated along "nonintuitive" Newtonian lines? And does not the claim that classical concepts are close to ordinary experience disregard the tremendous intellectual effort of the founders of modern science, who replaced the obvious and intuitive Aristotelian world with the counterintuitive Galilean-Newtonian one?

If so, the real question is why Bohr considered this controversial point to be the undebatable truth. In my view, Bohr's scientific style, shaped by years of his ingenious application of the correspondence principle, his sympathy with the British way of doing physics, his mistrust of abstract mathematics, and the cautious, reluctant quality of his radicalism — all these contributed to his belief in the indispensability of a classical framework. The a priori applicability of classical concepts obviously has a Kantian ring. Bohr probably absorbed Kantian presupposition from his philosophy teacher Höffding, whose lectures he attended, and the latter devoted a considerable part of his lectures to an elaboration of the Kantian system (Höffding 1906). Moreover, it seems that Höffding influenced Bohr's philosophical approach to problems in general (Jammer 1966, Holton 1970, and especially Wise 1989).

Important in this respect was also the influence of great English physicists whom Bohr admired. After finishing his Ph.D. degree, Bohr went to England to study with Thompson and Rutherford. Thompson preached the indispensability of forming pictures, or models, of abstract mathematical description (Heilbron 1977, 42), while Rutherford clearly depreciated the role of mathematics in deciphering the ways of Nature (Robertson 1979, 57). Bohr's admiration for Maxwell's achievement shines through much of his published work. Klein remembers Bohr discussing early physicists, like Maxwell and Raylegh, whom he talked of with great admiration, in 1918.[21] Bohr also learned of the Anglo-Saxon way of life through his father, who seemed to have had "an interest in everything English" (Robertson 1979). Bohr's acceptance of the British belief that forming pictures of mathematical descriptions of nature is an essential aspect of a physicist's activity is revealed in his 1917 proposal for a new institute of theoretical physics:

> In order that education in theoretical physics can be conducted in a manner corresponding to the subject's importance in the study of science, it is necessary that under guidance the student have the opportunity to carry out calculations and, through the demonstration and independent study of practical models, to become familiar with and form for themselves a clear picture of the often abstract and mathematically complicated theories of fundamental physical phenomena (quoted in Robertson 1979, 21).

Bohr's natural predisposition to pictures and classical models was considerably strengthened by his skillful use of the correspondence principle. The practical need for the continuity of physicists' tools was transformed into the doctrine of the indispensability of classical concepts. The correspondence principle had already split one reality, one world, into two halves — the classical and the quantum. The complementarity principle only legalized this paradoxical split: an inherently non-classical world must necessarily be described by classical concepts. Since this one world was already divided in two, why not introduce an additional necessary duality — that of waves and particles?

Neither Einstein nor Schrödinger, nor, for that matter, Heisenberg, liked this duality. They all preferred one unified system of concepts rather than two mutually exclusive ones. On this point Heisenberg agreed with Schrödinger who disliked the philosophy of *als ob* (as if). To treat the same object as if it were in one situation a particle and in another a wave is merely a juristic escape.[22] Heisenberg's complaint to Pauli echoed Schrödinger's dissatisfaction: if one assumes the existence of both waves and particles, everything can be made to appear contradiction-free.[23] Not indifferent to the economical beauty and power of abstract mathematics, Heisenberg, as he revealed in an interview with Kuhn,[24] never liked this dual approach to Nature. Probably, more than Bohr, he was aware of the limits of the kind of analysis in which his Danish colleague was so unusually gifted. When imprecise analogies became too disturbing to Heisenberg's aesthetic sensibility, he preferred to find solace in the haughty and soothing clarity of mathematical formulas:

> One may say that the concept of complementarity introduced by Bohr ... has encouraged the physicists to use an ambiguous rather than unambiguous language, to use the classical concepts in a somewhat vague manner ... always conscious of the fact that these concepts have only a very limited range of applicability. When this vague and unsystematic use of language leads into difficulties, the physicist has to withdraw into the mathematical scheme.... (Heisenberg 1959, 154.)

This sort of escape was not open to Bohr; he was, in Heisenberg's words, not a mathematical physicist, but a natural philosopher, never quite at ease with the esoteric game of symbols that his younger colleagues had mastered, it seemed, so effortlessly.

Till the end of his life Bohr struggled with one of the biggest cosmic riddles — how to make the "essentially" irrational quantum theory rationally comprehensible without recourse to mathematical abstraction. And if Einstein's last wish was: "If only I knew more mathematics," Bohr's fantasy was to write a book about quantum mechanics which would express all the important results without any mathematics at all (Holton 1970).

The new quantum mechanics, and especially the uncertainty principle, simply implied that the particle picture was bankrupt; both Einstein and Schrödinger took this position (Przibram 1961, 33). Concepts that have such a "shaky validity" simply must be given up, and some new conceptual framework, different from the "tranquilizing" orthodox version of quantum philosophy, should be sought (ibid.). Schrödinger did not lean towards wave concepts because he considered "waves more beautiful than particles," as Pauli insinuated; no more did Einstein prefer field concepts because he was a "hard-boiled determinist." Wave concepts simply seemed to both Schrödinger and Einstein less deficient in the quantum domain than the particle concept.

More than two decades after the complementarity principle was first announced, Schrödinger still believed in the necessity of a radical overhaul of physics, much more radical than the one that had occurred. The classical concepts of position, velocity and acceleration "as instantaneous properties of anything real" had worked for the entire era of classical physics, yet "now it seems to work no longer" (ibid., 38). Perhaps one should "go back 300 years and reflect on how one could have proceeded differently at the time, and how the whole subsequent development would then be modified" (ibid.).

Bohr had a more pragmatic turn of mind. What determines the evolution of physics, just like the evolution of phenomena in physics, is not only laws but given boundary conditions. And once the initial boundary conditions — those of the conceptual basis of modern science — are given, there is no way to turn the wheel of history backwards. Classical physics meant not only a true description of nature within the limits of its validity; it meant the availability of conceptual tools that cannot be "undone." The special status of classical concepts is inevitable not on logical but on historical grounds. Bohr's complementarity principle is thus not only a principle of the interpretation of physics, but also a principle of the growth of scientific knowledge.

In addition, the doctrine of the indispensability of classical concepts played an important ideological role. Combined with the belief in the finality of quantum theory and the completeness of its mathematical structure, it implied that the Copenhagen-Göttingen interpretation was not just one of the possible interpretations but the only conceivable one. Both doctrines — the finality of quantum theory and the indispensability of classical concepts — were crystallised in response to Schrödinger's challenge. Schrödinger found the abstract formulas of matrix mechanics repelling, and never abandoned the ardent hope to relegate "quantum jumps" to the status of "Ptolemaic epicycles." Schrödinger was not ready to remain content with the present state of quantum theory, and he was willing to sacrifice classical concepts in order to preserve the classical ideal of explanation (MacKinnon 1982; Beller 1985b).

In response to Schrödinger's threat, the Göttingen-Copenhagen physicists hastened to acclaim quantum theory as being immune to modification. This doctrine of finality was expressed by both Born and Heisenberg in their presentation at Como: "We maintain that quantum mechanics is a complete theory; its basic physical and mathematical hypotheses are not further susceptible of modification" (quoted in Jammer 1974, 114).

In this strategy of countering the opposition, the doctrine of the indispensability of classical concepts had a very significant role to play. First, it implied that the quantum world is not to be rejected as weird and unfathomable, for the regular classical concepts, if used in the Bohr-Heisenberg way, are sufficient in order to comprehend all the peculiarities and irrationalities introduced by the existence of the quantum of action. The applicability of classical concepts to the quantum domain answered Schrödinger's accusations concerning the basic non-intuitability of the Göttingen-Copenhagen version of quantum mechanics. The importance of this strategic move was, it seems, fully appreciated by both Bohr and Heisenberg, who quickly combined forces and smoothed over their disagreements over the uncertainty paper: "All that mattered now was to represent the facts in such a way that despite their novelty they could be grasped and accepted by all physicists" (Heisenberg 1971, 79).

Most important of all, however, the indispensability of classical concepts meant that the Copenhagen-Göttingen interpretation of physics was irrefutable. If quantum theory is complete and final, and the available classical concepts are the only possible ones, both uncertainty and complementarity follow with almost ironclad necessity. And this, of course, meant that one should not waste one's effort searching for an alternative — whether Schrödinger, Einstein or a lesser, "naive" and "old-fashioned" physicist. Perhaps in its strongest form, this overpowering ideology of finalizing past achievement and preventing critical questioning was expressed by Bohr's faithful pupil Leon Rosenfeld as late as 1961:

> Young physicists are raising doubts about the correctness of the basic ideas of quantum mechanics, and try to do it better. These efforts are, I am afraid, rather futile, because they rest on a complete misunderstanding of the problem ... there is nothing more accidental in the emergence of complementarity logic than in the emergence of man himself as a product of organic evolution ... I may therefore conclude on an optimistic note by assuring the younger generation that the instruments of rational analysis we are handing over to them contribute to a further increase of our power of understanding (Rosenfeld 1961, 384).

A clearer statement of the ideological role of complementarity hardly seems possible.

Acknowledgements

I would like to thank the Edelstein Center for the History and Philosophy of Science, Technology and Medicine at the Hebrew University of Jerusalem, for financial assistance supporting research for this paper. The research was conducted partly in the Berkeley Office for History of Science and Technology, and I would like to thank the Office and Professor John Heilbron for their hospitality, as well as Yehuda Elkana and the Van Leer Jerusalem Institute for the use of its facilities.

My sincere gratitude goes also to Dr. James Fuchs for his helpful comments and to Tina Beller for her support.

Notes

1. This term is due to Y. Elkana (Elkana 1977), who sees such "critical dialogues" as the central driving force in the growth of knowledge.
2. I have recently completed the paper "The Birth of Bohr's Complementarity — the Context and the Dialogues," which further substantiates this point and modifies the interpretation of the wave-particle duality in the Como paper.
3. This summary is based upon Beller 1983, 1985a, 1985b.
4. Interview with Heisenberg, *Archive for the History of Quantum Physics*, American Philosophical Society, Philadelphia, hereafter cited as *AHQP*.
5. Born's expression in interview with Kuhn, *AHQP*.
6. Bohr's letter to Rutherford, dated 27 January 1926, *AHQP*.
7. Reprinted as Bohr 1926.
8. See, for example, Pauli's letter to Bohr, 17 November 1925, *Pauli's Correspondence*, p.258.
9. Heisenberg, of course, had his good reasons to do so, as I have argued in Beller 1985b.
10. Interview with O. Klein, *AHQP*.
11. Interview with Heisenberg, *AHQP*.
12. Interview with Klein, *AHQP*.
13. There are several reasons for Bohr's receptiveness, which I explore in my recent paper (see note 2).
14. Bohr to Kronig, 28 October 1926, in *AHQP*.
15. Most references in this paper are from Bohr's original presentation in Como (Bohr 1928a), which differs from the revised version published in *Nature* (Bohr 1928b).
16. Interview with Klein, *AHQP*.
17. Interview with Dirac, *AHQP*.
18. Interview with Klein, *AHQP*.
19. See James's description of the mental patient Lucie, whose consciousness was split into complementary parts that "coexist but mutually ignore each other," quoted in Jammer 1966.
20. Bohr to Schrödinger, 27 May 1928, *AHQP*.
21. Interview with Klein, *AHQP*.
22. Schrödinger to Bohr, 23 October 1926, *AHQP*.
23. Heisenberg to Pauli, 16 May 1927, in *Pauli's Correspondence*, p. 394.
24. Interview with Kuhn, *AHQP*.

References

Beller, M., 1983. "Matrix Theory Before Schrödinger — Problems, Philosophy, Consequences," *Isis* 74:469–91.

—, 1985a. "Pascual Jordan's Influence on the Discovery of Heisenberg's Indeterminacy Principle," *Archive for History of Exact Sciences* 33:337–49.

—, 1985b. "Epistemological Opportunism and the Rise of Quantum Mechanics," to appear in *Boston Studies for History and Philosophy of Science*.

—, 1986. "John Hendry. *The Creation of Quantum Mechanics and the Bohr-Pauli Dialogue*," *Isis* 77:107–9.

Bohr, N., 1926. "Atomic Theory and Mechanics," *Nature* 116.

—, 1928a. "The Recent Development in Atomic Theory and the Quantum Postulate," *Atti del Congresso Internazionale dei Fisici, Como, Settembre 1927*. Bologna: Zonichelli.

—, 1928b. "The Recent Development in Atomic Theory and the Quantum Postulate," *Nature* 121:580–90.

Bohr, N., H. Kramers, and T. Slater, 1924. "The Quantum Theory of Radiation," *Philosophical Magazine* 47:785–802.

Born, M., 1926a. "Zur Quantenmechanik der Stossvorgaenge," *Zeitschrift fur Physik* 37:863–67.

—, 1926b. "Quantenmechanik der Stossvorgänge," *Zeitschrift für Physik* 38:803–27.

Bothe, W. and M. Geiger, 1925. "Über das Wesen des Comptoneffects: ein experimenteller Beitrag zur Theorie der Strahlung," *Zeitschrift für Physik* 32:639–63.

Capek, M., 1961. *The Philosophical Impact of Contemporary Physics*. Princeton: Van Nostrand.

Elkana, Y., 1977. "Historical Roots of Modern Physics," in Weiner (ed.), 1977, 197–265.

Heilbron, J., 1977. "Lectures on the History of Atomic Physics," in Weiner (ed.), 1977.

—, 1988. "The Earliest Missionaries of the Copenhagen Spirit," in *Science in Reflection*, ed. E. Ullmann-Margalit. Dordrecht: Kluwer Academic Publishers, 201–33.

Heisenberg, W., 1925. "Quantentheoretische Umdeutung kinematischer und mechanischer Beziehungen," *Zeitschrift für Physik* 33:879–93.

—, 1927. "Über den anschaulichen Inhalt der Quantentheoretischen Kinematik und Mechanik," *Zeitschrift für Physik* 43:172–98.

—, 1959. *Physics and Philosophy*. London: Allen and Unwin.
—, 1971. *Physics and Beyond: Encounters and Conversations*. New York: Allen and Unwin.
Herman, A., K. V. Meyer, and V. F. Weisskopf (eds.), 1979. *Pauli's Scientific Correspondence*. New York: Springer-Verlag.
Höffding, H. 1906. *The Problems of Philosophy,* trans. M. Fisher. New York: Macmillan.
Holton, Gerald, 1970. "The Roots of Complementarity," *Daedalus*, Fall: 1015–55, reprinted in *Thematic Origins of Scientific Thought*. Cambridge, Mass.: Harvard University Press, 115–61.
Jammer, M., 1966. *The Conceptual Development of Quantum Mechanics*. New York: McGraw-Hill.
—, 1974. *The Philosophy of Quantum Mechanics: The Interpretations of Quantum Mechanics in Historical Perspective*. New York: Wiley.
Keller, E. F., 1979. "Cognitive Repression in Contemporary Physics," *American Journal of Physics* 478:718–21.
Kennard, E. H., 1928. "Note on Heisenberg's Indetermination Principle," *Physical Review* 31:344–48.
MacKinnon, E., 1982. *Scientific Explanation and Atomic Physics*. Chicago: University of Chicago Press.
Pauli's Correspondence. See Herman et al., 1979.
Przibram, K., 1967. *Letters on Wave Mechanics*. New York: Philosophical Library.
Robertson, P., 1979. *The Early Years*. Copenhagen: Akademishe Forlag.
Rosenfeld, L., 1961. "Foundations of Quantum Theory and Complementarity," *Nature* 190:384–88.
Slater, J., 1975. "Wave Mechanics in the Classical Decade, 1923–1932," in *Solid State and Molecular Theory: A Scientific Biography*. New York: Wiley.
Weiner, C. (ed.), 1977. *History of 20th Century Physics*. New York: Academic Press.
Wise, M. N., 1989. "How Do Sums Count? On the Cultured Origins of Statistical Causality," in *The Probabilistic Revolution,* ed. L. Kruger, G. Gigerenzer, and M. J. Morgan. Cambridge, Mass.: MIT Press.

THE BAR-HILLEL COLLOQUIUM:
STUDIES IN HISTORY, PHILOSOPHY AND SOCIOLOGY OF SCIENCE

Volume 1
THE KALEIDOSCOPE OF SCIENCE
Edited by EDNA ULLMANN-MARGALIT

BOSTON STUDIES IN THE PHILOSOPHY OF SCIENCE 94
ISBN 90-277-2158-0 (HB)
ISBN 90-277-2159-9 (PB)

Volume 2
THE PRISM OF SCIENCE
Edited by EDNA ULLMANN-MARGALIT

BOSTON STUDIES IN THE PHILOSOPHY OF SCIENCE 95
ISBN 90-277-2160-2 (HB)
ISBN 90-277-2161-0 (PB)

Volume 3
SCIENCE IN REFLECTION
Edited by EDNA ULLMANN-MARGALIT

BOSTON STUDIES IN THE PHILOSOPHY OF SCIENCE 110
ISBN 90-277-2712-0 (HB)
ISBN 90-277-2713-9 (PB)

45. A. Ishimoto (ed.): Japanese Studies in the History and Philosophy of Science. (In prep.) ISBN 90-277-0733-3
46. P.L. Kapitza: *Experiment, Theory, Practice.* Articles and Addresses. Edited by R.S. Cohen. 1980 ISBN 90-277-1061-9; Pb 90-277-1062-7
47. M.L. Dalla Chiara (ed.): *Italian Studies in the Philosophy of Science.* 1981 ISBN 90-277-0735-9; Pb 90-277-1073-2
48. M.W. Wartofsky: *Models.* Representation and the Scientific Understanding. [Synthese Library 129] 1979 ISBN 90-277-0736-7; Pb 90-277-0947-5
49. T.D. Thao: *Phenomenology and Dialectical Materialism.* Edited by R.S. Cohen. 1986 ISBN 90-277-0737-5
50. Y. Fried and J. Agassi: *Paranoia.* A Study in Diagnosis. [Synthese Library 102] 1976 ISBN 90-277-0704-9; Pb 90-277-0705-7
51. K.H. Wolff: *Surrender and Cath.* Experience and Inquiry Today. [Synthese Library 105] 1976 ISBN 90-277-0758-8; Pb 90-277-0765-0
52. K. Kosík: *Dialectics of the Concrete.* A Study on Problems of Man and World. 1976 ISBN 90-277-0761-8; Pb 90-277-0764-2
53. N. Goodman: *The Structure of Appearance.* [Synthese Library 107] 1977 ISBN 90-277-0773-1; Pb 90-277-0774-X
54. H.A. Simon: *Models of Discovery* and Other Topics in the Methods of Science. [Synthese Library 114] 1977 ISBN 90-277-0812-6; Pb 90-277-0858-4
55. M. Lazerowitz: *The Language of Philosophy.* Freud and Wittgenstein. [Synthese Library 117] 1977 ISBN 90-277-0826-6; Pb 90-277-0862-2
56. T. Nickles (ed.): *Scientific Discovery, Logic, and Rationality.* 1980 ISBN 90-277-1069-4; Pb 90-277-1070-8
57. J. Margolis: *Persons and Mind.* The Prospects of Nonreductive Materialism. [Synthese Library 121] 1978 ISBN 90-277-0854-1; Pb 90-277-0863-0
58. G. Radnitzky and G. Andersson (eds.): *Progress and Rationality in Science.* [Synthese Library 125] 1978 ISBN 90-277-0921-1; Pb 90-277-0922-X
59. G. Radnitzky and G. Andersson (eds.): *The Structure and Development of Science.* [Synthese Library 136] 1979 ISBN 90-277-0994-7; Pb 90-277-0995-5
60. T. Nickles (ed.): *Scientific Discovery.* Case Studies. 1980 ISBN 90-277-1092-9; Pb 90-277-1093-7
61. M.A. Finocchiaro: *Galileo and the Art of Reasoning.* Rhetorical Foundation of Logic and Scientific Method. 1980 ISBN 90-277-1094-5; Pb 90-277-1095-3
62. W.A. Wallace: *Prelude to Galileo.* Essays on Medieval and 16th-Century Sources of Galileo's Thought. 1981 ISBN 90-277-1215-8; Pb 90-277-1216-6
63. F. Rapp: *Analytical Philosophy of Technology.* Translated from German. 1981 ISBN 90-277-1221-2; Pb 90-277-1222-0
64. R.S. Cohen and M.W. Wartofsky (eds.): *Hegel and the Sciences.* 1984 ISBN 90-277-0726-X
65. J. Agassi: *Science and Society.* Studies in the Sociology of Science. 1981 ISBN 90-277-1244-1; Pb 90-277-1245-X

Boston Studies in the Philosophy of Science

66. L. Tondl: *Problems of Semantics.* A Contribution to the Analysis of the Language of Science. Translated from Czech. 1981
ISBN 90-277-0148-2; Pb 90-277-0316-7
67. J. Agassi and R.S. Cohen (eds.): *Scientific Philosophy Today.* Essays in Honor of Mario Bunge. 1982 ISBN 90-277-1262-X; Pb 90-277-1263-8
68. W. Krajewski (ed.): *Polish Essays in the Philosophy of the Natural Sciences.* Translated from Polish and edited by R.S. Cohen and C.R. Fawcett. 1982
ISBN 90-277-1286-7; Pb 90-277-1287-5
69. J.H. Fetzer: *Scientific Knowledge.* Causation, Explanation and Corroboration. 1981 ISBN 90-277-1335-9; Pb 90-277-1336-7
70. S. Grossberg: *Studies of Mind and Brain.* Neural Principles of Learning, Perception, Development, Cognition, and Motor Control. 1982
ISBN 90-277-1359-6; Pb 90-277-1360-X
71. R.S. Cohen and M.W. Wartofsky (eds.): *Epistemology, Methodology, and the Social Sciences.* 1983. ISBN 90-277-1454-1
72. K. Berka: *Measurement.* Its Concepts, Theories and Problems. Translated from Czech. 1983 ISBN 90-277-1416-9
73. G.L. Pandit: *The Structure and Growth of Scientific Knowledge.* A Study in the Methodology of Epistemic Appraisal. 1983 ISBN 90-277-1434-7
74. A.A. Zinov'ev: *Logical Physics.* Translated from Russian. Edited by R.S. Cohen. 1983 ISBN 90-277-0734-0
See also Volume 9.
75. G-G. Granger: *Formal Thought and the Sciences of Man.* Translated from French. With and Introduction by A. Rosenberg. 1983 ISBN 90-277-1524-6
76. R.S. Cohen and L. Laudan (eds.): *Physics, Philosophy and Psychoanalysis.* Essays in Honor of Adolf Grünbaum. 1983 ISBN 90-277-1533-5
77. G. Böhme, W. van den Daele, R. Hohlfeld, W. Krohn and W. Schäfer: *Finalization in Science.* The Social Orientation of Scientific Progress. Translated from German. Edited by W. Schäfer. 1983 ISBN 90-277-1549-1
78. D. Shapere: *Reason and the Search for Knowledge.* Investigations in the Philosophy of Science. 1984 ISBN 90-277-1551-3; Pb 90-277-1641-2
79. G. Andersson (ed.): *Rationality in Science and Politics.* Translated from German. 1984 ISBN 90-277-1575-0; Pb 90-277-1953-5
80. P.T. Durbin and F. Rapp (eds.): *Philosophy and Technology.* [*Also* Philosophy and Technology Series, Vol. 1] 1983 ISBN 90-277-1576-9
81. M. Marković: *Dialectical Theory of Meaning.* Translated from Serbo-Croat. 1984 ISBN 90-277-1596-3
82. R.S. Cohen and M.W. Wartofsky (eds.): *Physical Sciences and History of Physics.* 1984. ISBN 90-277-1615-3
83. É. Meyerson: *The Relativistic Deduction.* Epistemological Implications of the Theory of Relativity. Translated from French. With a Review by Albert Einstein and an Introduction by Milić Čapek. 1985 ISBN 90-277-1699-4

Boston Studies in the Philosophy of Science

84. R.S. Cohen and M.W. Wartofsky (eds.): *Methodology, Metaphysics and the History of Science.* In Memory of Benjamin Nelson. 1984 ISBN 90-277-1711-7
85. G. Tamás: *The Logic of Categories.* Translated from Hungarian. Edited by R.S. Cohen. 1986 ISBN 90-277-1742-7
86. S.L. de C. Fernandes: *Foundations of Objective Knowledge.* The Relations of Popper's Theory of Knowledge to That of Kant. 1985 ISBN 90-277-1809-1
87. R.S. Cohen and T. Schnelle (eds.): *Cognition and Fact.* Materials on Ludwik Fleck. 1986 ISBN 90-277-1902-0
88. G. Freudenthal: *Atom and Individual in the Age of Newton.* On the Genesis of the Mechanistic World View. Translated from German. 1986 ISBN 90-277-1905-5
89. A. Donagan, A.N. Perovich Jr and M.V. Wedin (eds.): *Human Nature and Natural Knowledge.* Essays presented to Marjorie Grene on the Occasion of Her 75th Birthday. 1986 ISBN 90-277-1974-8
90. C. Mitcham and A. Hunning (eds.): *Philosophy and Technology II.* Information Technology and Computers in Theory and Practice. [*Also* Philosophy and Technology Series, Vol. 2] 1986 ISBN 90-277-1975-6
91. M. Grene and D. Nails (eds.): *Spinoza and the Sciences.* 1986 ISBN 90-277-1976-4
92. S.P. Turner: *The Search for a Methodology of Social Science.* Durkheim, Weber, and the 19th-Century Problem of Cause, Probability, and Action. 1986. ISBN 90-277-2067-3
93. I.C. Jarvie: *Thinking about Society.* Theory and Practice. 1986 ISBN 90-277-2068-1
94. E. Ullmann-Margalit (ed.): *The Kaleidoscope of Science.* The Israel Colloquium: Studies in History, Philosophy, and Sociology of Science, Vol. I. 1986 ISBN 90-277-2158-0; Pb 90-277-2159-9
95. E. Ullmann-Margalit (ed.): *The Prism of Science.* The Israel Colloquium: Studies in History, Philosophy, and Sociology of Science, Vol. II. 1986 ISBN 90-277-2160-2; Pb 90-277-2161-0
96. G. Márkus: *Language and Production.* A Critique of the Paradigms. Translated from French. 1986 ISBN 90-277-2169-6
97. F. Amrine, F.J. Zucker and H. Wheeler (eds.): *Goethe and the Sciences: A Reappraisal.* 1987 ISBN 90-277-2265-X; Pb 90-277-2400-8
98. J.C. Pitt and M. Pera (eds.): *Rational Changes in Science.* Essays on Scientific Reasoning. Translated from Italian. 1987 ISBN 90-277-2417-2
99. O. Costa de Beauregard: *Time, the Physical Magnitude.* 1987 ISBN 90-277-2444-X
100. A. Shimony and D. Nails (eds.): *Naturalistic Epistemology.* A Symposium of Two Decades. 1987 ISBN 90-277-2337-0
101. N. Rotenstreich: *Time and Meaning in History.* 1987 ISBN 90-277-2467-9
102. D.B. Zilberman: *The Birth of Meaning in Hindu Thought.* Edited by R.S. Cohen. 1988 ISBN 90-277-2497-0

103. T.F. Glick (ed.): *The Comparative Reception of Relativity*. 1987
ISBN 90-277-2498-9
104. Z. Harris, M. Gottfried, T. Ryckman, P. Mattick Jr, A. Daladier, T.N. Harris and S. Harris: *The Form of Information in Science*. Analysis of an Immunology Sublanguage. With a Preface by Hilary Putnam. 1989 ISBN 90-277-2516-0
105. F. Burwick (ed.): *Approaches to Organic Form*. Permutations in Science and Culture. 1987 ISBN 90-277-2541-1
106. M. Almási: *The Philosophy of Appearances*. Translated from Hungarian. 1989
ISBN 90-277-2150-5
107. S. Hook, W.L. O'Neill and R. O'Toole (eds.): *Philosophy, History and Social Action*. Essays in Honor of Lewis Feuer. With an Autobiographical Essay by L. Feuer. 1988 ISBN 90-277-2644-2
108. I. Hronszky, M. Fehér and B. Dajka: *Scientific Knowledge Socialized*. Selected Proceedings of the 5th Joint International Conference on the History and Philosophy of Science organized by the IUHPS (Veszprém, Hungary, 1984). 1988 ISBN 90-277-2284-6
109. P. Tillers and E.D. Green (eds.): *Probability and Inference in the Law of Evidence*. The Uses and Limits of Bayesianism. 1988 ISBN 90-277-2689-2
110. E. Ullmann-Margalit (ed.): *Science in Reflection*. The Israel Colloquium: Studies in History, Philosophy, and Sociology of Science, Vol. III. 1988
ISBN 90-277-2712-0; Pb 90-277-2713-9
111. K. Gavroglu, Y. Goudaroulis and P. Nicolacopoulos (eds.): *Imre Lakatos and Theories of Scientific Change*. 1989 ISBN 90-277-2766-X
112. B. Glassner and J.D. Moreno (eds.): *The Qualitative-Quantitative Distinction in the Social Sciences*. 1989 ISBN 90-277-2829-1
113. K. Arens: *Structures of Knowing*. Psychologies of the 19th Century. 1989
ISBN 0-7923-0009-2
114. A. Janik: *Style, Politics and the Future of Philosophy*. 1989
ISBN 0-7923-0056-4
115. F. Amrine (ed.): *Literature and Science as Modes of Expression*. With an Introduction by S. Weininger. 1989 ISBN 0-7923-0133-1
116. J.R. Brown and J. Mittelstrass (eds.): *An Intimate Relation*. Studies in the History and Philosophy of Science. Presented to Robert E. Butts on His 60th Birthday. 1989 ISBN 0-7923-0169-2
117. F. D'Agostino and I.C. Jarvie (eds.): *Freedom and Rationality*. Essays in Honor of John Watkins. 1989 ISBN 0-7923-0264-8
118. D. Zolo: *Reflexive Epistemology*. The Philosophical Legacy of Otto Neurath. 1989 ISBN 0-7923-0320-2
119. M. Kearn, B.S. Philips and R.S. Cohen (eds.): *Georg Simmel and Contemporary Sociology*. 1989 ISBN 0-7923-0407-1
120. T.H. Levere and W.R. Shea (eds.): *Nature, Experiment and the Science*. Essays on Galileo and the Nature of Science. In Honour of Stillman Drake. 1989
ISBN 0-7923-0420-9

Boston Studies in the Philosophy of Science

121. P. Nicolacopoulos (ed.): *Greek Studies in the Philosophy and History of Science*. 1990 ISBN 0-7923-0717-8
122. R. Cooke and D. Costantini (eds.): *Statistics in Science*. The Foundations of Statistical Methods in Biology, Physics and Economics. 1990 ISBN 0-7923-0797-6
123. P. Duhem: *The Origins of Statics*. Translated from French by G.F. Leneaux, V.N. Vagliente and G.H. Wagner. With an Introduction by S.L. Jaki. 1991 ISBN 0-7923-0898-0
124. H. Kamerlingh Onnes: *Through Measurement to Knowledge*. The Selected Papers, 1853-1926. Edited and with an Introduction by K. Gavroglu and Y. Goudaroulis. 1991 ISBN 0-7923-0825-5
125. M. Čapek: *The New Aspects of Time: Its Continuity and Novelties*. Selected Papers in the Philosophy of Science. 1991 ISBN 0-7923-0911-1
126. S. Unguru (ed.): *Physics, Cosmology and Astronomy, 1300-1700*. Tension and Accommodation. 1991 ISBN 0-7923-1022-5
127. Z. Bechler: *Newton's Physics on the Conceptual Structure of the Scientific Revolution*. 1991 ISBN 0-7923-1054-3
128. É. Meyerson: *Explanation in the Sciences*. Translated from French by M-A. Siple and D.A. Siple. 1991 ISBN 0-7923-1129-9
129. A.I. Tauber (ed.): *Organism and the Origins of Self*. 1991 ISBN 0-7923-1185-X
130. F.J. Varela and J-P. Dupuy (eds.): *Understanding Origins*. Contemporary Views on the Origin of Life, Mind and Society. 1992 ISBN 0-7923-1251-1
131. G.L. Pandit: *Methodological Variance*. Essays in Epistemological Ontology and the Methodology of Science. 1991 ISBN 0-7923-1263-5
132. G. Munévar (ed.): *Beyond Reason*. Essays on the Philosophy of Paul Feyerabend. 1991 ISBN 0-7923-1272-4
133. T.E. Uebel (ed.): *Rediscovering the Forgotten Vienna Circle*. Austrian Studies on Otto Neurath and the Vienna Circle. Partly translated from German. 1991 ISBN 0-7923-1276-7
134. W.R. Woodward and R.S. Cohen (eds.): *World Views and Scientific Discipline Formation*. Science Studies in the [former] German Democratic Republic. Partly translated from German by W.R. Woodward. 1991 ISBN 0-7923-1286-4
135. P. Zambelli: *The Speculum Astronomiae and Its Enigma*. Astrology, Theology and Science in Albertus Magnus and His Contemporaries. 1992 ISBN 0-7923-1380-1
136. P. Petitjean, C. Jami and A.M. Moulin (eds.): *Science and Empires*. Historical Studies about Scientific Development and European Expansion. ISBN 0-7923-1518-9
137. W.A. Wallace: *Galileo's Logic of Discovery and Proof*. The Background, Content, and Use of His Appropriated Treatises on Aristotle's *Posterior Analytics*. 1992 ISBN 0-7923-1577-4

138. W.A. Wallace: *Galileo's Logical Treatises*. A Translation, with Notes and Commentary, of His Appropriated Latin Questions on Aristotle's *Posterior Analytics*. 1992 ISBN 0-7923-1578-2
Set (137 + 138) ISBN 0-7923-1579-0
139. M.J. Nye, J.L. Richards and R.H. Stuewer (eds.), *The Invention of Physical Science*. Intersections of Mathematics, Theology and Natural Philosophy since the Seventeenth Century. Essays in Honor of Erwin N. Hiebert. 1992
ISBN 0-7923-1753-X
140. G. Corsi, M.L. dalla Chiara and G.C. Ghirardi (eds.), *Bridging the Gap: Philosophy, Mathematics and Physics*. Lectures on the Foundations of Science. 1992 ISBN 0-7923-1761-0
141. C.-H. Lin and D. Fu (eds.), *Philosophy and Conceptual History of Science in Taiwan*. 1992 ISBN 0-7923-1766-1
142. S. Sarkar (ed.), *The Founders of Evolutionary Genetics*. A Centenary Reappraisal. 1992 ISBN 0-7923-1777-7
143. J. Blackmore (ed.), *Ernst Mach – A Deeper Look*. Documents and New Perspectives. 1992 ISBN 0-7923-1853-6
144. P. Kroes and M. Bakker (eds.), *Technological Development and Science in the Industrial Age*. New Perspectives on the Science–Technology Relationship. 1992 ISBN 0-7923-1898-6
145. S. Amsterdamski, *Between History and Method*. Disputes about the Rationality of Science. 1992 ISBN 0-7923-1941-9
146. E. Ullmann-Margalit (ed.), *The Scientific Enterprise*. The Bar-Hillel Colloquium: Studies in History, Philosophy, and Sociology of Science, Volume 4. 1992 ISBN 0-7923-1992-3

Also of interest:
R.S. Cohen and M.W. Wartofsky (eds.): *A Portrait of Twenty-Five Years Boston Colloquia for the Philosophy of Science, 1960-1985*. 1985 ISBN Pb 90-277-1971-3

Previous volumes are still available.

KLUWER ACADEMIC PUBLISHERS – DORDRECHT / BOSTON / LONDON

FSC
www.fsc.org
MIX
Papier | Fördert
gute Waldnutzung
FSC® C083411

Kolibri 360 GmbH
Ferdinand-Jühlke-Straße 7
99095 Erfurt, Deutschland
produktsicherheit@kolibri360.de